Catalyst Poisoning

CHEMICAL INDUSTRIES

A Series of Reference Books and Text Books

Consulting Editor
HEINZ HEINEMANN
*Heinz Heinemann, Inc.,
Berkeley, California*

Volume 1: Fluid Catalytic Cracking with Zeolite Catalysts, *Paul B. Venuto and E. Thomas Habib, Jr.*

Volume 2: Ethylene: Keystone to the Petrochemical Industry, *Ludwig Kniel, Olaf Winter, and Karl Stork*

Volume 3: The Chemistry and Technology of Petroleum, *James G. Speight*

Volume 4: The Desulfurization of Heavy Oils and Residua, *James G. Speight*

Volume 5: Catalysis of Organic Reactions, *edited by William R. Moser*

Volume 6: Acetylene-Based Chemicals from Coal and Other Natural Resources, *Robert J. Tedeschi*

Volume 7: Chemically Resistant Masonry, *Walter Lee Sheppard, Jr.*

Volume 8: Compressors and Expanders: Selection and Application for the Process Industry, *Heinz P. Bloch, Joseph A. Cameron, Frank M. Danowski, Jr., Ralph James, Jr., Judson S. Swearingen, and Marilyn E. Weightman*

Volume 9: Metering Pumps: Selection and Application, *James P. Poynton*

Volume 10: Hydrocarbons from Methanol,
Clarence D. Chang

Volume 11: Foam Flotation: Theory and Applications, *Ann N. Clarke and David J. Wilson*

Volume 12: The Chemistry and Technology of Coal, *James G. Speight*

Volume 13: Pneumatic and Hydraulic Conveying of Solids,
O. A. Williams

Volume 14: Catalyst Manufacture: Laboratory and Commercial Preparations, *Alvin B. Stiles*

Volume 15: Characterization of Heterogeneous Catalysts,
edited by Francis Delannay

Volume 16: BASIC Programs for Chemical Engineering Design,
James H. Weber

Volume 17: Catalyst Poisoning,
L. Louis Hegedus and Robert W. McCabe

Additional Volumes in Preparation

Catalyst Poisoning

L. Louis Hegedus
Research Division
W. R. Grace & Co.
Columbia, Maryland

Robert W. McCabe
General Motors Research Laboratories
Warren, Michigan

MARCEL DEKKER, INC. New York and Basel

Library of Congress Cataloging in Publication Data

Hegedus, L. Louis, [Date]
   Catalyst poisoning.

   (Chemical Industries ; 17)
   Bibliography: p.
   Includes index.
   1. Catalyst poisoning. I. McCabe, Robert W.,
[Date]   II. Title. III. Series.
TP156.C35H44   1984   660.2'995   84-11405
ISBN 0-8247-7173-7

The contents of this book originally appeared in Catalysis Reviews - Science and Engineering, Volume 23, Number 3, 1981 edited by Heinz Heinemann, published by Marcel Dekker, Inc., New York.

COPYRIGHT © 1984 by MARCEL DEKKER, INC. ALL RIGHTS RESERVED

Neither this book nor any part may be reproduced or transmitted in any form or by any means, electronic or mechanical, including photocopying, microfilming, and recording, or by any information storage and retrieval system, without permission in writing from the publisher.

MARCEL DEKKER, INC.
270 Madison Avenue, New York, New York 10016

Current printing (last digit):
10 9 8 7 6 5 4 3 2 1

PRINTED IN THE UNITED STATES OF AMERICA

All things are poison and nothing is without a poison, the dose alone makes a thing not a poison.

Paracelsus

Preface

Catalytic processes represent the mainstream of chemical technologies today. Worldwide catalyst sales amount to about $2.2 billion per year (1982), and the economic impact of catalysis reaches by perhaps two orders of magnitude beyond this figure.

The durability of catalysts is one of the most important of their characteristics, since it determines, to a significant extent, the economics of their use.

Despite the technical and economical importance of catalyst poisoning, no book has, at the time of this writing, yet been specifically devoted to the subject, even though several review articles exist. Thus, the publication of this short book may fill a gap in the catalytic literature.

The book's material has its origins in a review prepared for the International Symposium on Catalyst Deactivation in Antwerp, Belgium, October 1980. Several reviews were presented at that meeting; our task was to discuss the chemical deactivation (i.e., poisoning) of catalysts. The Proceedings were published in B. Delmon and G. F. Froment (eds.), Catalyst Deactivation, Elsevier, Amsterdam, 1980.

An extended version of our above-mentioned review paper appeared more recently in Catalysis Reviews—Science and Engineering 23(3), 377 (1981). Marcel Dekker, Inc., publisher of the above journal, suggested that we publish the material in the form of this book; the text was updated and slightly extended for this purpose.

L. Louis Hegedus

Contents

Preface		v
1. Introduction		1
2. Mechanism and Kinetics		5
	A. Introductory Remarks	5
	B. Monofunctional Catalysts with Uniform Sites	5
	C. Monofunctional Catalysts with Site Strength Distribution	28
	D. Multifunctional Catalysts	34
	E. Support Effects	43
3. Intra- and Interparticle Transport Effects		47
	A. Introductory Remarks	47
	B. Mechanistic Considerations	47
	C. Time-Dependent Behavior	56
	D. Effects of the Size and Shape of the Catalyst Pellets	64
	E. Pore Structure Effects	66
	F. Effects of Catalyst Impregnation Profiles	67
	G. Selectivity Problems	73
	H. Nonisothermal Catalyst Pellets	78
4. Design of Poison-Resistant Catalysts: A Case History		81
References		99
Index		109

Catalyst Poisoning

1
Introduction

The literature of catalysis is vast, reflecting the industrial importance of the subject. For the sake of curiosity, we undertook a computerized search of Chemical Abstracts to trace the number of publications per year, using the generic form "Cataly-," and, for catalyst poisoning, the simultaneous occurrence of the generic forms "Cataly-" and "Poison-," beginning with the year 1967 and ending with 1978.

Our findings are displayed in Fig. 1 where the numbers of publications per year in the above two categories were normalized against their values in 1967.

Both catalysis and catalyst poisoning appear to be mature fields, with cumulated average annual growth rates of 4.8 and 5.2%, respectively. To our surprise, however, less than 1% of the publications in catalysis are formally devoted to catalyst poisoning. While some implications of such a cursory search may be questionable, the conclusion is nevertheless clear: catalyst poisoning does not seem to receive adequate attention in the catalytic literature, especially, as we found, on the fundamental level.

For practical purposes, three main classes of catalyst deactivation can be distinguished: chemical, thermal, and mechanical.

This review concentrates on chemical deactivation, involving a chemically induced change in the catalyst's activity. This change may be related to the competitive, reversible adsorption of the poison precursor (inhibition); to the irreversible adsorption, deposition, or reaction of the poison precursor on or with the surface (poisoning); to the poison-induced restructuring of the surface; or to the physical blockage of the support's pore structure (pore plugging).

1

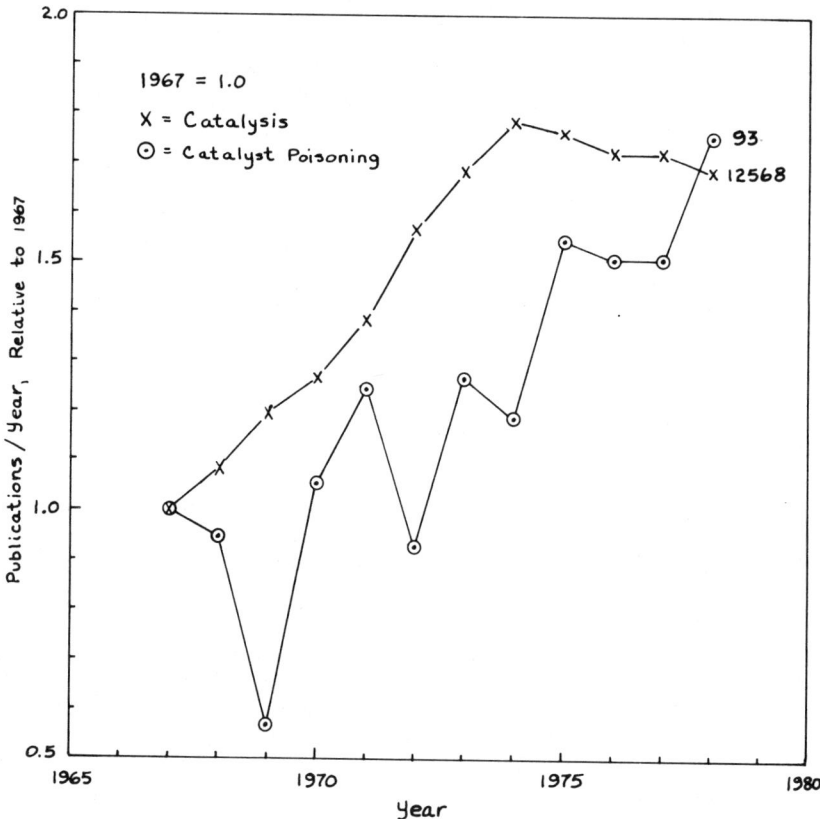

FIG. 1. Results of a computerized scanning of Chemical Abstracts.

According to the precursor of the poisoning event, we distinguish impurity poisoning from self-poisoning; in the latter case, one or more of the reaction participants (reactants, products, intermediates) serve as poison precursors. A large class of self-poisoning reactions is represented by catalyst coking and we will touch upon that area only lightly.

Thermal deactivation is often difficult to separate from chemical deactivation, since catalyst sintering, restructuring, alloying, alloy segregation, metal volatilization, and various thermally induced metal-support interactions often show a strong sensitivity toward the chemical environment of the catalyst. In this review we will quote such effects only to the extent that they are chemically induced.

This review will be centered around gas-solid systems. The emphasis will be on metal or metal oxide catalysts, while zeolites will not be considered.

Instead of claiming an exhaustive search of the literature, we would like to refer to several excellent reviews by others (e.g., Berkman et al. [1], Maxted [2], Innes [3], Petro [4], Butt [5], Petro [6], Knozinger [7], Butt and Billimoria [8], Oudar [9], and Butt [10]), and concentrate on a systematic discussion of chemical deactivation by quoting appropriate examples, mostly of the last decade. In discussing the various modes of chemical deactivation, we will progress from simpler toward more complex events, so that when citing more complex phenomena, we can refer to their building blocks previously described.

In writing this review we tried to avoid the breakdown of the material into surface science, catalytic chemistry, and engineering aspects. Instead, we attempted to discuss mechanism and kinetics simultaneously, at the level of detail which is necessary to describe the phenomenon involved. We hope that the engineer will benefit from appreciating the often intricate chemistry of the events which can lead to catalyst poisoning. In turn, by discussing the mathematical modeling of these phenomena, we hoped to expose the chemist to the engineering approach which aims at distinguishing and quantifying essential features.

We will begin our discussion with the poisoning of the catalytic surface and progress only to the scale of a catalyst pellet, discussing the role of transport effects within and around it. The effects of poisoning on the performance of catalytic reactors will be outside of our scope.

Working in an industrial environment, the authors felt that some benefit could be derived from describing a case history where attempts toward a fundamental understanding of complex catalyst deactivation events and their mathematical analysis contributed to the successful development of catalysts with improved poison resistance. The last section will describe such an example, taken from the authors' experience.

2
Mechanism and Kinetics

A. Introductory Remarks

Poisoning problems are composed of two components: the nature of the interactions of the poison with the catalytic surface, and the effects of these interactions on the reactions the catalyst is intended to promote. Since catalyst poisoning manifests itself in affecting these (often already complex) main reactions, catalyst poisoning studies face the cumulated complexity of the above two events.

In this part of our review we exploit the advantages of classifying the modes of chemical deactivation of catalysts according to the degree of complexity of the phenomena involved, in the sequence of monofunctional catalysts with monofunctional sites, monofunctional catalysts where site strength distribution is important, multifunctional catalysts, and catalysts where the nature of the support surface introduces phenomena which do not occur over nonsupported catalysts.

B. Monofunctional Catalysts with Uniform Sites

Although even ideal single crystal surfaces may show a significant complexity of catalytic sites, the simple concept of site uniformity can often be invoked to result in useful interpretation of catalytic events.

There are three main categories of catalyst poisoning which need to be distinguished at this point: <u>poison adsorption</u>, <u>poison-induced surface reconstruction</u>, and <u>compound formation</u> between the poison and the catalyst.

There are numerous ways in which adsorption of poisonous species can affect catalytic activity, but one of the most common mechanisms is through competitive adsorption with reactant species. Poison adsorption is often termed reversible or irreversible, where reversibility is either defined by the recovery of the activity upon removal of the poison from the feedstream under actual reaction conditions, or by the recovery of activity upon changing the feedstream or the operating conditions (regeneration).

The reversibility of poison adsorption can be conveniently quantified in thermodynamic terms. Thus McCarty and Wise [11] measured sulfur chemisorption isosteres on both powdered and alumina-supported Ni catalysts. Mixtures of H_2S and H_2 were exposed to the catalyst, and the sulfur coverages were calculated from measuring the composition of the gas phase after equilibration by the reaction

$$H_2S(g) \rightleftharpoons H_2(g) + S(a) \qquad (1)$$

with

$$P_{H_2S}/P_{H_2} = K_p = \exp\left(\frac{\Delta H}{RT} - \frac{\Delta S}{R}\right) \qquad (2)$$

where ΔH and ΔS are referenced to gaseous H_2S and H_2. The isosteres for $Ni/\alpha\text{-}Al_2O_3$ are shown in Fig. 2. For temperatures in the range of 400 to 900 K, sulfur coverages of half of the saturation coverage or larger were observed for H_2S/H_2 partial pressure ratios as low as 1 ppb, indicating that the equilibrium is strongly shifted toward a sulfur-covered surface.

Figure 3 shows the heats of adsorption calculated from the isosteres. According to the authors, the discontinuity above $\xi = 1.12$ probably represents a change in adsorbing species, from S adatoms to adsorbed HS or H_2S. The adsorption isotherms could not be interpreted by Langmuirian adsorption, apparently due to a repulsive interaction between adjacent chemisorbed sulfur atoms. The energetics of sulfur interaction with Ni (Fig. 4) indicates that chemisorbed sulfur is more stable than the sulfur in Ni_3S_2 and sulfur dissolved in the bulk [S(b)/Ni]. This latter finding suggests a very large surface coverage of sulfur in equilibrium with very low levels of sulfur dissolved in the bulk.

Alstrup et al [12] conducted sulfur chemisorption isobar measurements over a $Ni/MgAl_2O_4$ catalyst in the range of 773 to 1023 K and $H_2S:H_2$ ratio of 7 to 50 ppm. They were successful in fitting the data using a model with a coverage-dependent heat of adsorption. The same model also well described the high-temperature data of previous investigators.

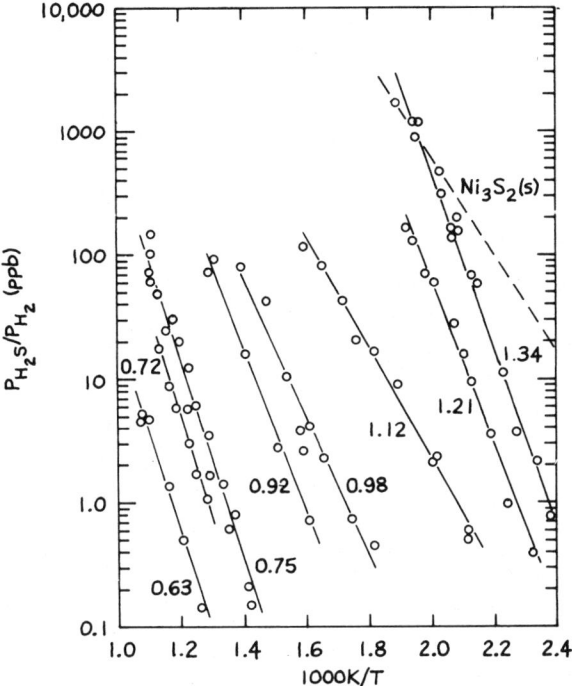

FIG. 2. Sulfur chemisorption isosteres on Ni/α-Al$_2$O$_3$. Parameter: coverage normalized to CO uptake at 300 K [11].

Benard et al. [13] conducted a systematic study of the thermodynamics of sulfur adsorption on a number of metals (Ag, Cu, Ni, Cr) and found an excellent correlation between the heat of adsorption of sulfur and the heat of formation of the corresponding bulk metal sulfides.

Even if all sites are taken to be equivalent in their energetics and geometry, the relationship between reaction rate and fraction of surface poisoned may still be quite complicated. Herington and Rideal [14] investigated the problem of multisite chemisorption of the reactant and the poison. Although they assumed that the poison is merely blocking a geometrically fixed number of sites, they point out that in practice a chemical interaction between the poison and the catalyst may result in electronic effects on adjacent sites with an action radius larger than the geometric proportions of the poison. In addition, they

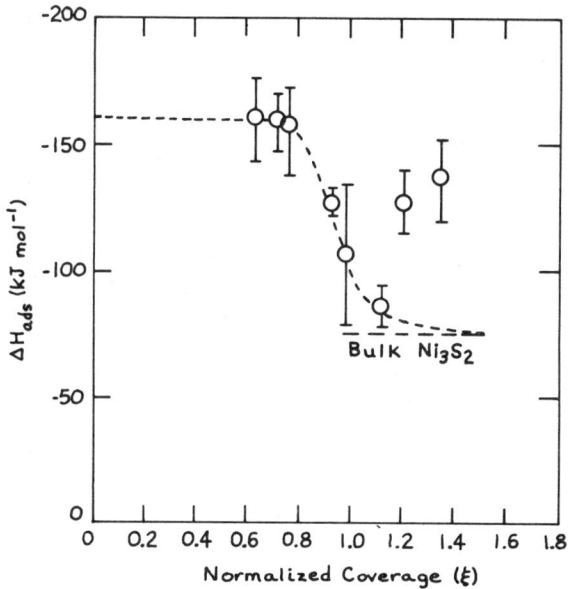

FIG. 3. Heat of adsorption of sulfur on Ni/α-Al$_2$O$_3$ as a function of coverage [11].

point out that the poison may even modify the electronic structure of the metallic crystal lattice; we will quote examples for both of such phenomena later on.

Two cases of poison-reactant interference were distinguished by Herington and Rideal [14], shown in Fig. 5. In Case a, the sets of adsorption sites can overlap, while in Case b they are widely separated.

Case b represents a situation where no mutual interference is possible, and thus the activity is proportional to the number of groups of active sites. For the case where the poison occupies one active site (a fraction θ_A of the active sites are bare) and an isolated group of n centers is required for reaction, the number of unpoisoned groups will be proportional to θ_A^n.

The more general situation is represented by Case a, where the sets of adsorption sites are overlapping.

Herington and Rideal [14] investigated the cases where the number of adsorption sites per reactant molecule can be m = 1, 2, or 7. For poison molecules it was either assumed that they take up n = 1 or

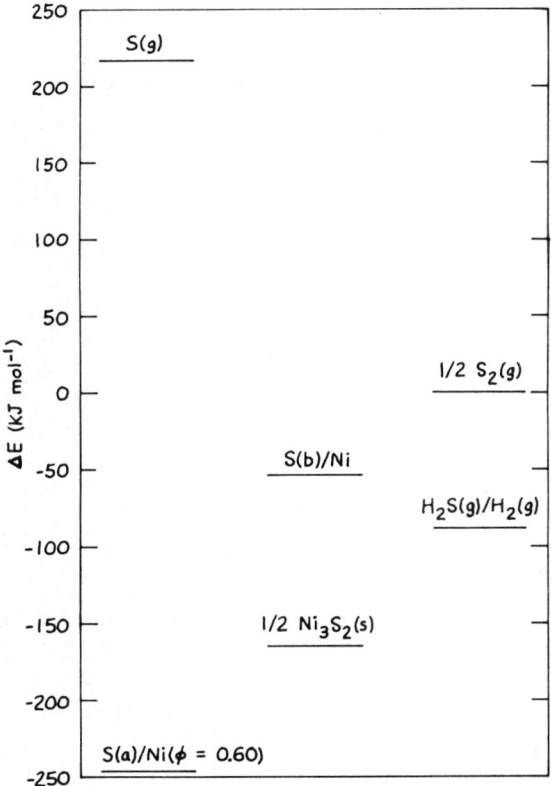

FIG. 4. Energetics of S interaction with Ni at 800 K [11].

7 sites, or that a single-site poison molecule is large enough to prevent poison adsorption (but not reactant adsorption) over its neighboring sites. A Monte Carlo technique was employed and the (111) plane of a face-centered cubic crystal was taken as an example.

For single-site (n = 1) poison adsorption, Fig. 6 shows the fraction of surface covered by the reactant as a function of the fraction of surface poisoned for m = 1, 2, and 7. The conclusion is that if the reactant requires more than one site for adsorption, the reaction is more readily poisoned.

For the case of a single-site but large poison molecule which cancels poison adsorption on neighboring sites, Fig. 7 shows that

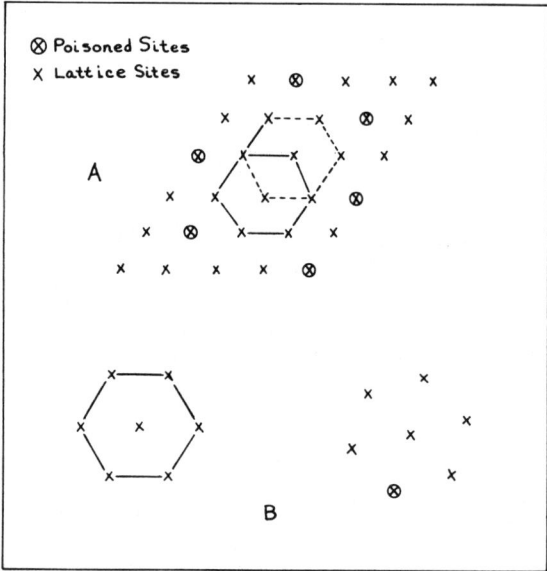

FIG. 5. A: Interference between two possible modes of adsorption. B: Adsorption on isolated sets of sites [14].

the surface cannot be completely poisoned if m = 1 or 2, but it can be completely poisoned for m = 7.

These calculations were carried out for the case where the surface structure of the adsorbed molecules is "frozen," i.e., no redistribution occurs. At a stationary state during reaction, as the authors explain, redistribution takes place since the reacted molecules will leave the surface, creating new sites for incoming reactant molecules. This leaves the above qualitative conclusions intact but will change the computed numerical values for reactant coverages.

The multisite competitive adsorption during surface reaction has been recently analyzed in more detail (for a poison-free case) by Frennet et al. [15]; it seems that the analysis could be extended to catalyst poisoning with multiple sites.

For two-site poisoning mechanisms with 1-, 2-, 3-, 4-, and 7-site reactant adsorption, the decline of available surface area with increasing fractional surface coverage by the poison was computed by Verma and Ruthven [16], using techniques similar to those of Herington and Rideal [14].

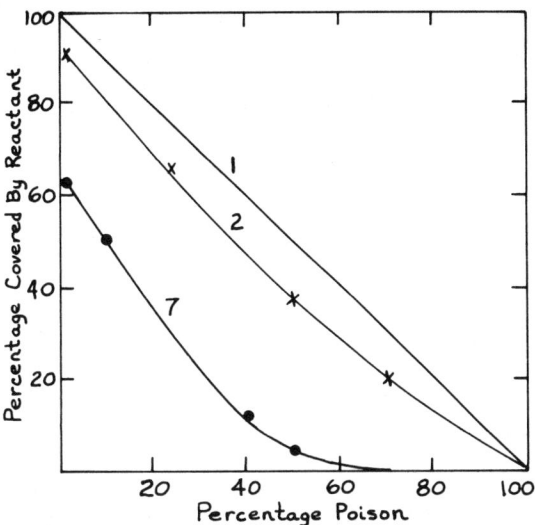

FIG. 6. Poisoning of a face-centered cubic (111) plane by a single-site poison for 1-, 2-, and 7-site reactants [14].

Multi-site poison chemisorption has indeed been often observed experimentally. Thus, Schwarz [17], while investigating the adsorption and desorption kinetics of hydrogen over clean and sulfur-covered Ru(001) surfaces, concluded that dissociative hydrogen chemisorption is completely inhibited at a sulfur coverage of 0.25, indicating that one sulfur atom poisons four Ru atoms for this reaction.

Similarly, Agrawal et al. [18] found that sulfur (from H_2S as a precursor) poisons Co/Al_2O_3 for CO hydrogenation to methane primarily due to a geometric blockage of sites, with one sulfur atom adsorbed per two surface Co atoms. However, electronic effects during sulfur adsorption were also found to be important, in that the activation energy of the methanation reaction was reduced upon partial poisoning.

The advent of modern surface-sensitive tools [low energy electron diffraction (LEED), Auger electron spectroscopy (AES), etc.] has allowed the detailed analysis of surface structures during catalyst poisoning. On Pt(100) surfaces, for example, sulfur is adsorbed in two well-defined structures (Berthier et al. [19], Heegemann et al. [20]), as shown in Fig. 8, taken from Fischer and Kelemen [21]. Figure 8(a) depicts the centered c(2X2) structure with a half-mono-

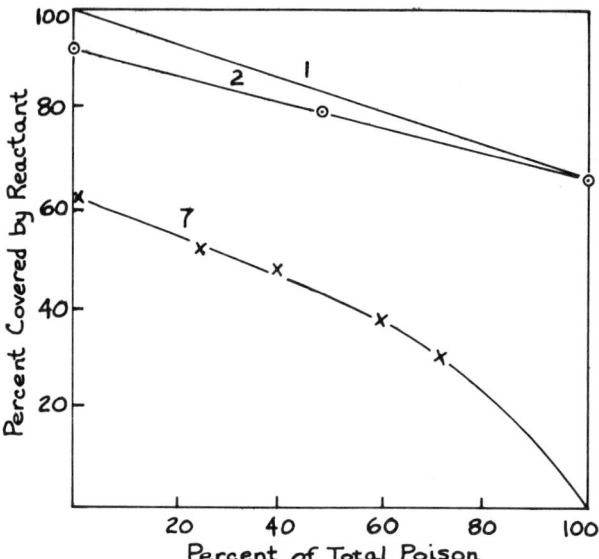

FIG. 7. Poisoning of a face-centered cubic (111) plane by a poison molecule too large to be adsorbed on adjacent sites, but the reactant may adsorb on sites adjacent to the poison. 1-, 2-, and 7-site reactants [14].

layer of sulfur, while a primitive p(2X2) structure (Fig. 8b) can also be formed containing one-quarter of a monolayer of sulfur. The c(2X2) structure is the highest sulfur coverage compatible with the van der Waals radius of sulfur for all ordered structures with respect to the Pt atoms. Similar structural observations have been carried out for a variety of other systems; an excellent example is the detailed analysis of sulfur structures over Ni(110) which change, with increasing exposures, through c(2X2), p(3X2), to p(5X2) (see, e.g., Mroz [22] and the recent review by Oudar [9]).

Beyond structural information about the chemisorbed poison, surface-sensitive techniques can also yield valuable information about the competition between adsorbing poisons and reactants. This competition can be highly selective: for example, Erley and Wagner [23] showed by LEED and thermal desorption spectroscopy (TDS) that in the low sulfur coverage regime ($\theta < 0.1$), one sulfur atom may block approximately nine potential sites for CO adsorption over Ni(111). Above a sulfur coverage of $\theta = 0.33$, CO adsorption stops completely.

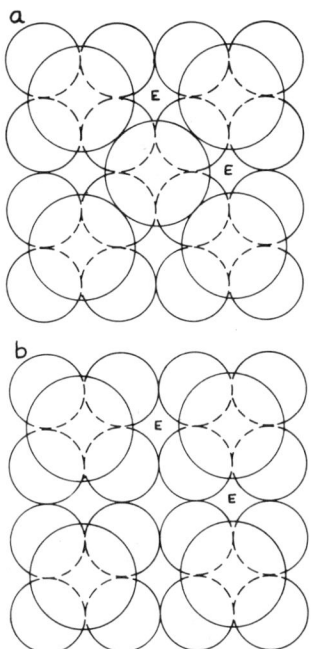

FIG. 8. Sulfur overlayers on a Pt(100) surface. a: Centered c(2X2) structure. b: Primitive p(2X2) structure [21].

Complex LEED patterns were shown to result from a coincidence structure of a chemisorbed sulfur overlayer.

Bonzel and Ku [24] investigated the interactions of CO and H_2S over Pt(110). Figure 9 shows flash desorption spectra (from saturation) of CO from partially sulfur-covered Pt(110). The most significant feature is the decrease in CO coverage (proportional to the area under the thermal desorption spectra) with increasing sulfur coverage, providing direct evidence for the competition between CO and S for the same sites. Sulfur preferentially poisons the high-energy binding states of CO, as indicated by the shift in the CO desorption peaks toward lower temperatures, concomitant with the development of additional features not observed in the sulfur-free system.

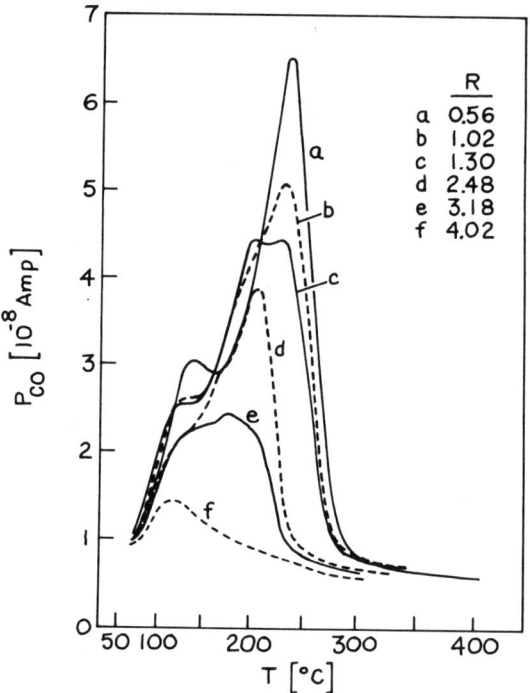

FIG. 9. Flash desorption of CO from partially sulfur-covered Pt(110). R refers to the relative sulfur concentration [24].

Figure 10 is a plot of the area under the CO flash desorption spectra as a function of sulfur coverage. At low coverages the data are consistent with each sulfur atom blocking two CO chemisorption sites. If this relationship were preserved at higher coverages, all CO chemisorption sites would be taken up at a sulfur coverage of $\theta = 0.5$. However, as Fig. 10 shows, at higher S coverages more CO is chemisorbed than the dual-site S chemisorption model would predict. This change in slope is due to phase changes in the adsorbed sulfur layer which allow more CO to be adsorbed than a simple c(2X2) sulfur structure would permit. LEED data indicate that these phase changes are related to the repulsive interactions between S atoms at higher coverages. Repulsive interactions also occur between S and CO, resulting in a lower-energy binding state for CO on Pt(110).

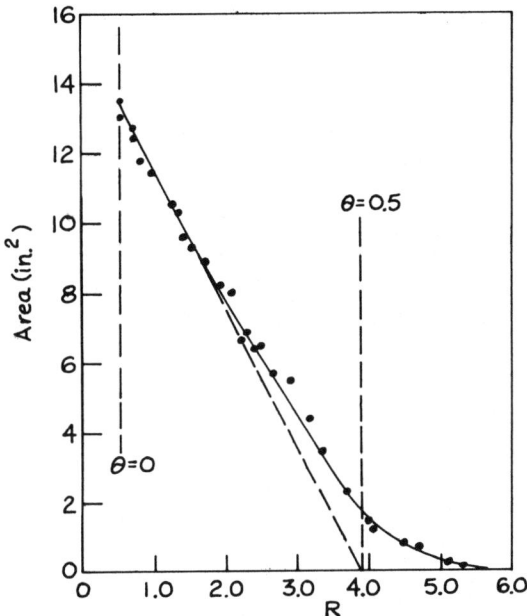

FIG. 10. Area under the CO flash desorption curves as a function of relative sulfur concentration R [24].

Bonzel and Ku [25] extended their studies to the effects of H_2S on the oxidation of CO over Pt(110). The interaction between sulfur and oxygen on this surface was found to be strong (resulting in reaction between the two). This is in contrast to the relatively weak interaction between surface-sulfur and surface-CO (Bonzel and Ku [24]) which primarily changes the chemisorption properties of CO. Reaction between sulfur and oxygen on the surface resulted in the formation of islands of sulfur and oxygen (determined by scanning AES). The size of these islands was dependent on the initial sulfur coverage, which determined the initial concentration of the free sites for oxygen adsorption and dissociation.

For low sulfur coverages ($\theta < 0.3$), the sulfur patches were found to be small and, for modeling purposes, the sulfur adlayer can be assumed to be uniform. At higher sulfur coverages ($\theta > 0.3$), large patches of sulfur are formed, and the surface is completely poisoned for the CO oxidation reaction.

Noting that the rate of sulfur oxidation is much slower than the rate of CO oxidation, the CO oxidation rate was modeled by an Eley-Rideal mechanism which neglects the rate of sulfur oxidation:

$$d[CO_2]/dt = kP_{CO}[O_{ad}] \qquad (3)$$

with

$$d[O_{ad}]/dt = k_{ad}P_{O_2}[\sigma]^2 - kP_{CO}[O_{ad}] \qquad (4)$$

the difference between the rates of oxygen adsorption and reaction (σ is the concentration of the free sites).

At steady state,

$$d[O_{ad}]/dt = 0 \qquad (5)$$

and thus

$$[O_{ad}] = \frac{k_{ad}P_{O_2}}{kP_{CO}} \sigma^2 \qquad (6)$$

Since

$$\sigma + \kappa[S_{ad}] = 1 \qquad (7)$$

(κ is a constant relating surface sulfur concentration to oxygen chemisorption sites), we obtain

$$[O_{ad}] = \frac{k_{ad}P_{O_2}}{kP_{CO}} (1 - \kappa[S_{ad}])^2 \qquad (8)$$

which, when substituted into Eq. (3), yields

$$d[CO_2]/dt = k_{ad}P_{O_2}(1 - \kappa[S_{ad}])^2 \qquad (9)$$

Equation (9) can be normalized against the clean surface rate ($[S_{ad}] = 0$)

$$d[CO_2]/dt = k_{ad}P_{O_2} \qquad (10)$$

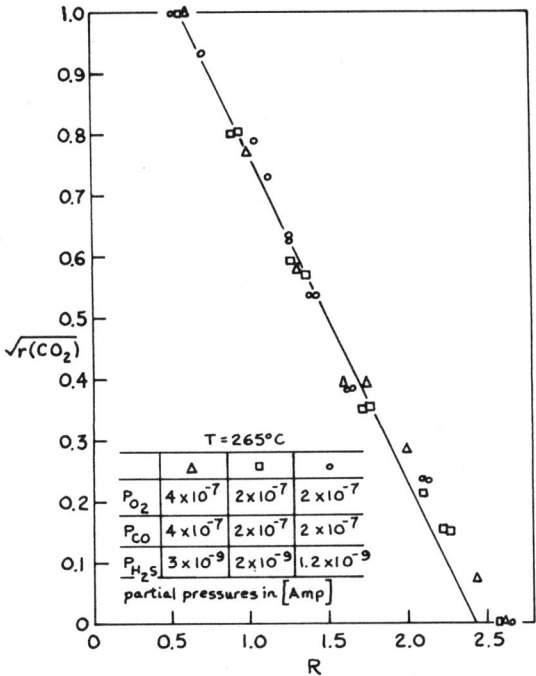

FIG. 11. Rate of CO_2 formation versus sulfur coverage [25].

Thus the relative rate of CO_2 production is

$$r_{CO_2} = (1 - \kappa [S_{ad}])^2 \qquad (11)$$

Figure 11 shows the relationship between the square root of r_{CO_2} and a quantity R which is proportional to $[S_{ad}]$, showing excellent linearity.

In another example for modeling the chemistry of catalyst poisoning, the kinetics of self-poisoning reactions has been analyzed by Wolf and Petersen [26], employing Langmuir-Hinshelwood rate expressions for both the main and the poisoning reactions. If there is a rate-determining step in the reaction sequence and if the rate of the poisoning reaction is slow, the steady-state assumption allows the derivation of the roots of several empirical rate laws. The value of such detailed

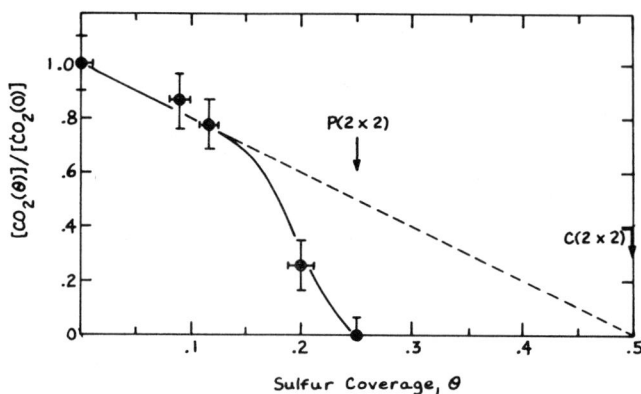

FIG. 12. Rate of the CO + NO reaction as a function of sulfur coverage on Pt(100) [21].

analyses is that they incorporate mechanistic information into manageable rate expressions, thus enhancing their utility.

The sulfur poisoning of a Pt(100) surface was analyzed by Fischer and Kelemen [21] for the reduction of NO by CO and for the dehydrogenation of benzene and acetylene. Figure 12 shows the relative rate of CO_2 formation as a function of sulfur surface coverage for the reaction

$$2CO + 2NO \rightarrow 2CO_2 + N_2 \tag{12}$$

At high sulfur coverages the surface is covered by a centered c(2X2) saturation sulfur layer ($\theta = 0.5$) and the surface is catalytically inactive.

At intermediate sulfur coverages a primitive p(2X2) structure prevails. The cage structure of this arrangement leaves an open site surrounded by four sulfur atoms. Additional sulfur atoms and gas molecules can be adsorbed in the middle of these sulfur squares, and the reactants are prevented from participating in a Langmuir-Hinshelwood-type surface reaction. Thus their surface mobility is inhibited by the poison which keeps them separated.

At all sulfur coverages below saturation, the strong chemical bond formed with the sulfur weakens the interaction of Pt with other adsorbates, most likely through a modification of the electronic properties of Pt in the vicinity of the S atom. We will discuss such phenomena in more detail later on.

In Fig. 12 the dashed line represents the reaction rate as a function of S coverage if all adsorbed molecules were able to react. The solid line resulted from a Monte Carlo simulation by considering a p(2X2) overlayer of sulfur with CO and NO adsorbed at random in the middle of the sulfur squares. Sulfur atoms were then removed at random and replaced with NO and CO molecules. The probability of having an island with at least two of each reagent in the island is 5/8, so 5/8th of the number of single sulfur vacancies represents the number of CO_2 molecules formed according to the stoichiometry of Reaction (12).

Figure 12 shows the excellent agreement of this "cage model" of sulfur poisoning with experimental data.

Another interesting example for the case of poison-weakened adsorbate-surface interaction (and resulting activity and selectivity changes) has been documented by Johnson and Madix [27]. The chemisorption and reactions of CH_3OH over clean Ni(100) and over two S-poisoned Ni(100) surfaces (p2X2)S, with 0.25 monolayer of S, and c(2X2)S, with 0.50 monolayer of S) were investigated, employing temperature-programmed reaction spectroscopy. Sulfur weakened the chemisorption of methanol over the Ni surface: over clean Ni(100) and over p(2X2)S, the primary reaction was decomposition to CO and H_2, while formaldehyde and H_2 were formed over c(2X2)S, via a methoxy (CH_3O) intermediate.

Poison-induced changes in bonding represent a fairly large class of poisoning problems. Reactant-catalyst bonds can be weakened, and we have discussed excellent examples of this effect, such as the weakening of the CO-metal bond by adsorbed sulfur (Fischer and Kelemen [21] and references therein, Bonzel and Ku [24], etc.). There is another kind of bond interference, however: the effect of poisons on the strength of the molecular bonds within the chemisorbed reactant molecules. Such an example is discussed by Rhodin and Brucker [28], who investigated the changes in the molecular bonding of chemisorbed CO on α-Fe(100) in the presence of sulfur.

UV photoemission spectroscopy (UPS) showed that when CO is chemisorbed over a sulfur-free surface, the C—O bond has a stretched configuration relative to gaseous CO, which makes it more likely to dissociate. However, the presence of sulfur reduces both the forward- and back-donation of electrons between CO and Fe, resulting in a relaxation of the CO molecule toward the unstretched (and more stable) configuration. The electron bonding interactions between CO and Fe can be quenched to varying degrees by chemisorbed C, O, P, or S, with S causing a completely unstretched configuration. Such effects on bonding caused by adsorbed sulfur have also been observed by infrared spectroscopy. As an example, Rochester and Terrell [29] found that the strength of the chemisorption bond between CO (linearly bonded) and Ni was weakened by sulfiding the metal.

Strong electronic effects have been claimed for the sulfur poisoning of the CO methanation reaction over Ni(100) by Goodman and Kiskinova [30]: one sulfur atom deactivated about ten Ni atom sites. The same authors (Kiskinova and Goodman, [31]) explored the effects of three electronegative poisons (P, S, Cl) on the chemisorption of CO and H_2 over Ni(100). The presence of these electronegative atoms on the surface reduced the adsorption rate, the adsorption bond strength, and the adsorption capacity of the surface, for both species. The extent of poisoning correlates well with the increasing electronegativity of the sequence P, S, Cl.

When comparing the data from single-crystal CO methanation studies (Goodman and Kiskinova [30]) with methanation studies on supported Ni catalysts (Rostrup-Nielsen and Pedersen [32]), it appears that the geometric effect prevails for higher sulfur coverages, while strong electronic effects prevail at low sulfur coverages over single-crystal Ni(100).

In addition to the papers we discussed above, there are numerous recent studies which deal with various aspects of poison adsorption. Sulfur-containing systems have been reviewed by Oudar [9] and investigated, among others, by Williams et al. [33], Kelemen et al. [34], Fisher [35], Chang [36], Windawi and Katzer [37], Cramb et al. [38], Bain et al. [39], Baird et al. [40], Otto et al. [41], Pannell et al. [42], Ng and Martin [43], Wood et al. [44], Rewick and Wise [45], Gonzalez-Tejuca and Turkevich [46], and Schwaha et al. [47]. The complexities of sulfur poisoning of the Fischer-Tropsch synthesis are discussed in the review of Madon and Shaw [48]. The utility of the pulse chromatographic method was illustrated by Yanovskii and Berman [49] on the example of the reversible poisoning (by pyridine) of cumene cracking over silica-alumina. Various aspects of sulfur, lead, and phosphorus poisoning can be found in the review of Shelef [50] and references therein, pertinent to automobile exhaust catalysis.

There are interesting complexities which may arise beyond those discussed above, and their details are often little understood. Thus poison precursors may react in the gas phase forming inert species; e.g., phosphorus and lead precursors in automobile exhaust (Acres et al. [51]). The addition of a component to the feedstream may suppress poisoning. For example, the toxicity of NH_3 toward the Pt-catalyzed hydrogenation of cyclohexene (Maxted and Biggs [52]) was suppressed by H_2O, and the toxicity of sulfur toward the reduction of NO by NH_3 over Pt, Ru, and Ni was suppressed by O_2 (Tsai et al.

[53]). On the other hand, the addition of H_2O to the feedstream during methanation over a supported Ni catalyst enhanced the poisoning effect of H_2S (Dalla Betta and Shelef [54]).

One poison may displace another poison, such as thionaphthene by thiophene over Pt (Maxted and Ball [55]), or a poison which is reversibly adsorbed at lower temperatures may form a poisonous species at higher temperatures which is irreversible; e.g., dimethylphenylarsine over Pt (Maxted and Ball [56]).

Adsorbed poisons may impart a selectivity change: a good example is the suppression of NH_3 formation over automobile exhaust catalysts by SO_2 (e.g., Trimble [57]). On Ni(100), sulfur impurities have little influence on the formation of carbidic intermediates during catalytic methanation, but the rate of hydrogenation is reduced (Madey et al. [58]). Poisons may impart selectivity by creating new active sites: an example is the lead poisoning of the selective hydrogenation of isoprene over Pd (Fuji and Bailar [59]).

Poisons may enhance the activity of the catalyst; e.g., the presence of Pb promotes the reduction of NO by CO over CuO (Sorensen and Nobe [60]), while it poisons the oxidation of ethylene (Sorensen and Nobe [61]). Undesirable side reactions may also be promoted by the deposition of poisons. For example, metals (e.g., Ni) found in crude oil deposit on fluidized cracking catalysts and promote excessive hydrogen and coke production. This can be suppressed by the addition of Sb which was shown to form an alloy with the deposited Ni, thereby both geometrically blocking the Ni surface and altering the electronic properties of the Ni atoms in such a way that their catalytic activity is reduced (Parks et al. [62]).

Often one of the reactants may have an influence on poisoning: examples can be taken from hydrogenation catalysis where the partial pressure of H_2 delays self-poisoning by coking (e.g., hydrodenitrogenation, see the review of Katzer and Sivasubramanian [63]).

An interesting case of self-poisoning was reported by Iida and Tamaru [64]: the exchange reaction between D_2 and H_2O over Pt was poisoned by a frozen water layer (below 273 K) which hindered the diffusion of deuterium to the Pt.

Depending on the operating conditions, a gas-phase impurity may act either as a promoter or a poison. Thus, during catalytic methanation over Ni(100), $Fe(CO)_5$ impurities in the feedstream act as promoters at T < 600 K but promote the formation of an inactive graphitic carbon surface layer at higher temperatures, thus causing poisoning (Madey et al. [58]).

Adsorbed poisons often may dissolve in the bulk phase of the catalyst in addition to poisoning the surface. Thus Williams and Baron

[65] showed by Auger electron spectroscopy that the differences in the susceptibilities of Pt and Pd to lead poisoning in oxidizing automobile exhaust are due to the fact that the lead penetrates the bulk of a Pd foil but remains on the surface of a Pt foil. Similar experiments by Tsai et al. [66] showed that sulfur incorporates into the bulk of Pt, Pd, and Ni foils but not into Ru during the reduction of NO by NH_3 in the presence of SO_2. As a reverse of this process, Kelley et al. [67] quote an example where sulfur may diffuse from the bulk of a Ni crystal to its surface causing its deactivation, apparently contributing to the formation of an inactive graphitic carbon layer during catalytic methanation.

The poison may form an alloy with the catalyst. For example, Affrossman et al. [68], studying the Hg poisoning of H_2-O_2 titration over Pt black, concluded that an alloy was formed between Pt and Hg, with Pt at the surface of the alloy retaining reactivity for the titration reaction. Further amounts of Hg may adsorb on top of this alloy layer, rendering it inactive.

Bulk dissolution of one of the reactants may also lead to self-poisoning. For example, Frackiewicz et al. [69] found that H_2 dissolved in the bulk of a Ni-Cu alloy changed the transition metal character of the catalyst by forming a β-Ni-Cu-H phase, resulting in a substantial decrease in activity for the rates of heterogeneous recombination of H atoms, para-ortho hydrogen conversion, and ethylene hydrogenation.

In addition to poison adsorption, chemically induced surface reconstruction may often significantly contribute to the state of activity of a catalyst, which may manifest itself in either increasing (activation) or decreasing (poisoning) the activity of the surface.

Schmidt and Luss [70] investigated the activation of Pt-10% Rh gauzes (commercially employed in the oxidation of NH_3 and in the production of HCN from NH_3, CH_4, and air) by 100 ppm H_2S, and showed that H_2S caused an entirely different surface morphology when compared with the catalyst before H_2S treatment. The surface was essentially free of sulfur after this high-temperature treatment, and consisted predominantly of (100) planes or another plane with two- or fourfold symmetry.

Sulfur poisoning of catalysts may also result in surface morphology changes. Somorjai [71] refers to several reported instances [e.g., the (111) face of Ni develops a (100) orientation in the presence of H_2S, C_2H_4, or benzene (McCaroll et al. [72]), or the (110) crystal face of Cr develops a (100) orientation in the presence of sulfur], and proposed that the presence of adsorbed impurities may alter the surface free energies of the crystal planes which will then rearrange to new planes with lower free energies. Thus the impurity causes a recrystallization of the surfaces. Surface free energies of various crystal surfaces have been computed by Herring [73], even though in the absence of impurities.

The extended Huckel method was employed by Halachev and Ruckenstein [74] to calculate the effects of electron acceptor elements (chemisorbed H, O, Cl, and S atoms) on the stability of PT(100) and PT(111) clusters, containing 9 and 10 atoms respectively. The adsorption of electron acceptors destabilizes the Pt clusters; the destabilizing effect decreases in the sequence $Cl \cong S > O > H$. This destabilization results in an increased mobility of the surface Pt atoms; the authors suggest that this is the first step in corrosive chemisorption and in the redispersion of supported Pt catalysts.

Catalytic etching during a reaction in the absence of poisons was reported by McCabe et al. [75] during NH_3 oxidation (with air) over Pt. Depending on the composition of the feedstream and on the temperature, different surface structures were observed. The "phase diagram" for this catalytic etching of Pt is displayed in Fig. 13, indicating a variety of complex structures.

The work of McCabe et al. [75] was continued by Flytzani-Stephanopoulos et al. [76], using small (~0.06 cm diameter) single crystal spheres of Pt, exposed to NH_3, C_3H_8, and CO oxidation, or NH_3 decomposition at high temperatures. Between one and five stable crystal planes were observed, and the planes formed and their relative areas depended both on the nature of the reaction and on the reactant composition. Minimum surface free energy considerations could account for unstable and metastable regions and also for faceting rates.

Amariglio and Rambeau [77] concluded that surface restructuring also must occur during NH_3 synthesis over Fe. Activation originates in an enhancement of surface defects, while spontaneous surface relaxation results in deactivation, they conclude.

The stability and structure of high Miller index Pt surfaces in various environments (vacuum, adsorbed carbon, O_2) was investigated by Blakely and Somorjai [78]. The surfaces restructured as the surface composition was changed, and the nature of the resulting structures depended on the nature of the adsorbed surface impurity.

Often the most serious form of catalyst poisoning is represented by compound formation between a poison precursor and the catalyst. There is no sharp boundary between compound formation and some of the stronger chemisorptive interactions discussed previously; as an example, during the sulfur poisoning of metallic catalysts, bulk metal sulfides may nucleate from two-dimensional sulfur overlayers, and such examples were already discussed (see, e.g., Fig. 3).

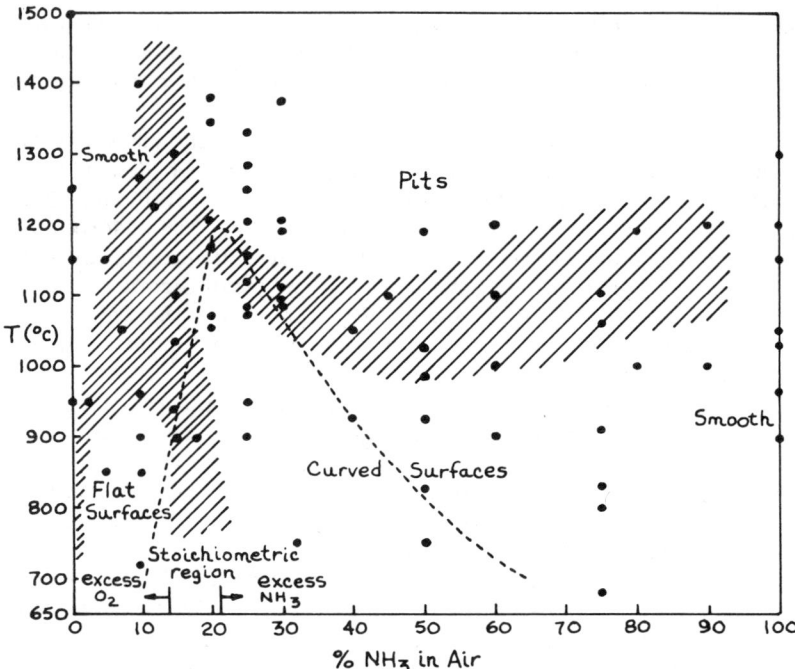

FIG. 13. Phase diagram for catalytic etching of Pt during NH_3 oxidation [75].

Compound formation may result both from impurity poisoning and self-poisoning, and we will discuss examples of both.

A prominent example of catalyst poisoning by compound formation with an impurity is represented by the sulfur poisoning of base metal catalysts in an oxidizing environment, such as in oxidizing automobile exhaust.

Fishel et al. [79] investigated the sulfur poisoning of Cu-Cr-Al_2O_3 catalysts in automobile exhaust. Figure 14 shows the detrimental effects of sulfur accumulation (in the form of sulfates) on CO emissions from a vehicle.

The stability regions of the Cu-S-O system were investigated by Ingraham [80]; his diagram, as quoted by Fishel et al. [79], is displayed in Fig. 15. Under the typical operating conditions of the data in Fig. 14, the phase diagram indicates that in an exhaust gas containing 10% O_2 and 10 ppm SO_2, $CuSO_4$ would decompose to CuO

UNIFORM SITES 25

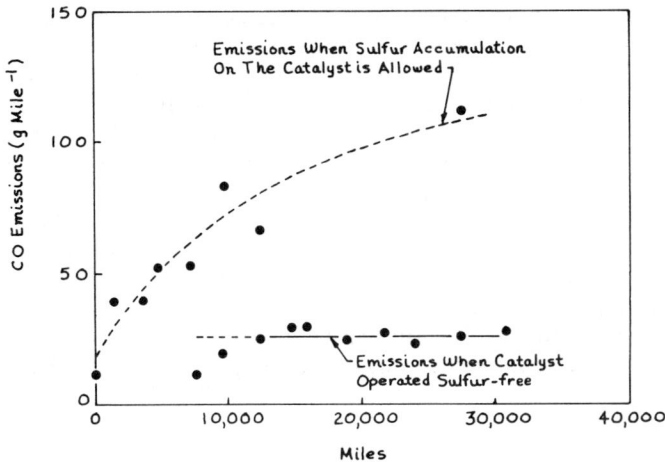

FIG. 14. Effects of sulfur on the performance of a base metal catalyst for automobile emission control [79].

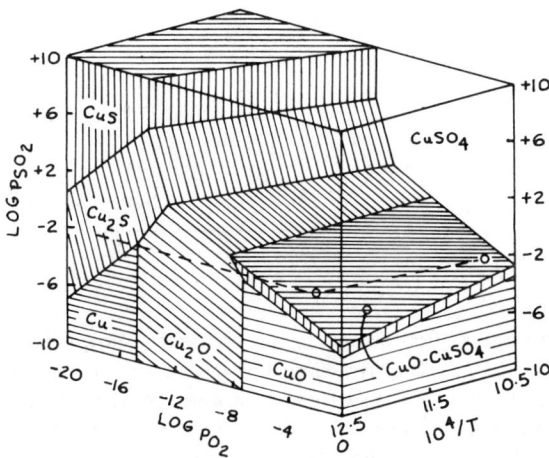

FIG. 15. Phase diagram of the Cu-S-O system (after Ref. 79).

if the temperature were held above 580 ± 15°C. In fact, experience has shown that sulfate poisoning of these catalysts can be effectively prevented if the temperature is held above about 600°C. Unfortunately, however, such high exhaust temperatures in vehicle exhaust can only be maintained at a significant fuel economy loss. The decomposition temperatures of a number of other sulfites and sulfates have been reported by Lowell et al. [81].

Base metal catalysts are sensitive to sulfur deactivation under reducing conditions as well, where stable sulfides may form. Simpson [82] conducted a thermodynamic analysis of the sulfur poisoning of prospective NO_x reduction catalysts in automobile exhaust. His calculated equilibrium desulfiding temperatures in the presence of 20 ppm H_2S in simulated automobile exhaust are shown in Table 1.

TABLE 1

Metal	Desulfiding temperature (°C)
Co	427
Cu	649 (Cu_2S)
Fe	649
Ir	371
Mn	None
Mo	538
Ni	371
Pb	482
Pd	371
Pt	371
Ru	482
Zn	None

The advantage of noble metals is apparent. However, some base metal catalysts (e.g., Co, Ni) have low-enough desulfiding temperatures to make them useful additives to noble metal systems.

Surface oxide formation also may poison the activity of catalysts. For example, Amirnazmi and Boudart [83] investigated the kinetics of NO decomposition over Pt in the temperature range of 600 to 1050°C. When O_2 was added to the NO feed, an oxidized surface was attained

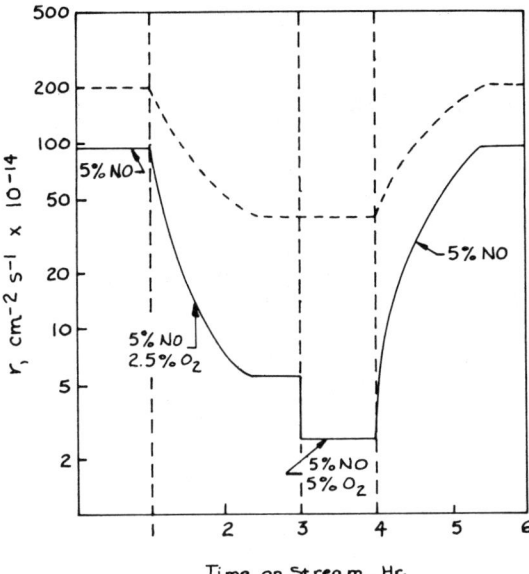

FIG. 16. Decomposition of NO on a Pt foil at 1000°C. Continuous line: Finite (short) contact time. Dotted line: Rate extrapolated to zero contact time [83].

which resulted in a lower reaction rate. The less active oxidized surface contains two oxygen atoms per Pt atom (PtO_2) and is produced by a slow rearrangement of the reduced Pt surface after O_2 was added. As Fig. 16 shows, the phenomenon is reversible: after removing the O_2 from the feed, the activity returns to its original, O_2-free level. Two modes of O_2 poisoning were distinguished: a slow reversible poisoning by surface-PtO_2 formation and an instantaneous reversible inhibition resulting from the competition for adsorption sites between NO and oxygen.

NO decomposition and the NO + CO reaction were studied over Fe and Ni films by Baker and Peterson [84]. The initial step was the decomposition of NO into N, O, and N_2O; the oxygen incorporated into the films to form an oxide layer (NiO or Fe_3O_4). The NO decomposition rate decreased as the thickness of this oxide layer increased; the films became completely poisoned when the oxide thickness exceeded 50 nm. Thus this is an example of self-poisoning resulting in compound formation.

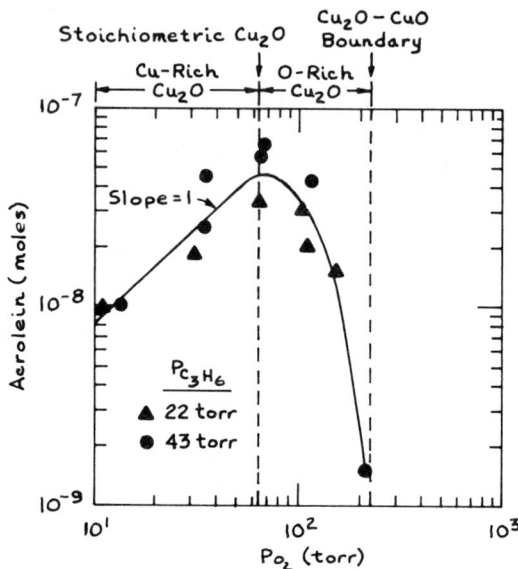

FIG. 17. Acrolein formation as a function of O_2 pressure over Cu_2O [86].

Surface oxides may also form on Rh, resulting, e.g., in the poisoning of CO oxidation (Kim et al. [85]).

Compound formation may also lead to interesting selectivity effects. Wood et al. [86], e.g., investigated the oxidation of propylene on the surface of a Cu_2O crystal according to the reaction

$$O_2 + \text{propylene} \to \text{acrolein} \to CO_2 + H_2O \tag{13}$$

where, of course, acrolein is the favored product. Stoichiometric or Cu-rich Cu_2O favors acrolein formation (Fig. 17) while oxygen-rich Cu_2O or CuO favors the complete oxidation to CO_2 and H_2O (Fig. 18).

C. Monofunctional Catalysts with Site Strength Distribution

The observation that the adsorption of often minute quantities of poisons may lead to the complete poisoning of certain catalytic reactions served as the prime motivator of the active site concept in catalysis (Taylor [87], Constable [88]), implying that only a small frac-

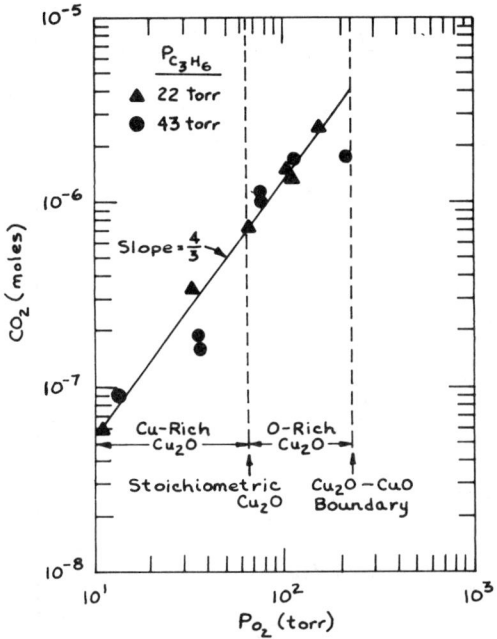

FIG. 18. Propylene oxidation to CO_2 over Cu_2O as a function of O_2 pressure [86].

tion of the catalytic surface (the "active sites") may be active for some reactions, and if these active sites are selectively covered by the poison, the activity of the catalyst declines to zero even if the total surface has not yet been covered by the poison.

We have seen in the previous parts of this review that similar phenomena can be envisioned even over perfectly uniform surfaces if multisite adsorption phenomena prevail (Herington and Rideal [14]), and will see examples later on which show that intrapellet diffusion resistances may also mimic such phenomena. Nevertheless, site nonuniformity has been conclusively proven to play a decisive role in the poisoning characteristics of numerous catalyst systems, and thus deserves discussion here.

If the site strengths are nonuniformly distributed, we can envision situations where the poison preferentially interacts with the most reactive sites (selective poisoning), with the least reactive sites (antiselective poisoning), or with all (nonselective poisoning), and ex-

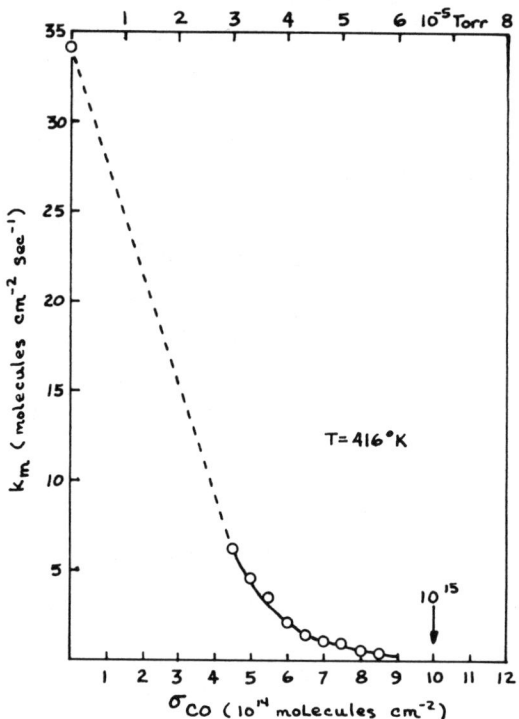

FIG. 19. CO poisoning of para-H_2 conversion over a Pt foil [89].

amples have been reported on all. We refer to this phenomenon as site strength selectivity.

The term selectivity can also be used in a different context: in the case of complex reactions where the different pathways are catalyzed by sites of different strengths, the poisoning of sites of a certain strength may alter the selectivity of the reaction system. We refer to this phenomenon as reaction selectivity, and will quote appropriate examples.

Selective poisoning is quite frequent, since the poisons often interfere preferentially with the most active sites. One example is provided by Volter and Hermann [89], who investigated the para-H_2 conversion reaction over a Pt foil in the presence of CO as a catalyst poison. Figure 19 shows the reaction rate as a function of CO cover-

SITE STRENGTH DISTRIBUTION

age of the surface. With increasing CO coverages, the first-order reaction rate coefficient decreased and the energy of activation of the reaction increased. The reaction is fastest over the sites with lower heats of H_2 adsorption, and these were preferentially poisoned by CO.

Schulz-Ekloff et al. [90] employed selective poisoning to investigate the crystal face specificity of NH_3 synthesis over WC grains under conditions free of diffusion resistances. Hexagonal WC crystals have two kinds of faces: a metallic and a carbidic one. The metallic face has been implicated to promote NH_3 synthesis. Selective H_2S poisoning of these metallic faces eliminated the catalyst's activity, thus reinforcing the above differentiation of the roles of metallic and carbidic faces of WC.

Baron [91] provides an interesting example of antiselective poisoning. The oxidation of CO over Pb-poisoned Pt films was investigated. The rate of CO oxidation was observed as a function of the fractional coverage of the surface by Pb (Fig. 20). The author suggested that at coverages <0.2 the lead preferentially binds to Pt sites which are less active for CO oxidation.

Clay and Petersen [92] studied the poisoning of an evaporated Pt film by AsH_3 during the reaction of cyclopropane hydrogenolysis. The activity of the catalyst declined linearly with the amount of AsH_3 adsorbed on the surface, thus exhibiting a characteristic nonselective behavior (Fig. 21). Both adsorption and reaction rate constants declined linearly with increasing poison coverage, indicating that the area for reaction is proportional to or the same as that for the adsorption of AsH_3.

A model was developed which invokes the sintering of Pt upon AsH_3 exposure, probably due to the formation of $PtAs_2$.

If a fraction Φ of the original Pt surface is poisoned by AsH_3, this fraction sinters to ϵ. Of this sintered area, a fraction $f\epsilon$ is active and $(1 - f)\epsilon$ is inactive, where

$$\epsilon/\Phi = \sigma \tag{14}$$

approximated by the ratio of surface roughness factors after and before the AsH_3-caused sintering.

The fraction of original surface area remaining is

$$\alpha = (1 - \Phi) + \epsilon \tag{15}$$

while the fraction of original activity remaining is

$$\eta = (1 - \Phi) + f\epsilon \tag{16}$$

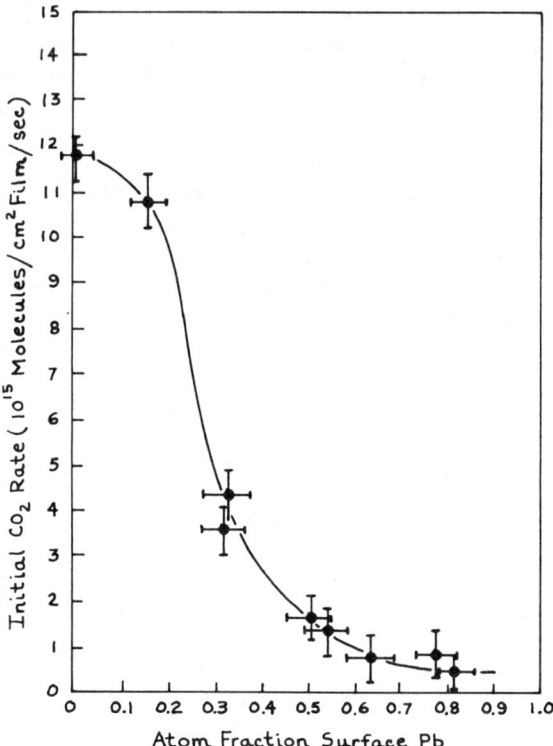

FIG. 20. Effect of lead coverage on the rate of CO oxidation over a Pt film [91].

Equations (14) to (16) can be combined to yield the experimentally observed linear relationship

$$\eta = 1 - (1 - \alpha)\left(\frac{1 - f\sigma}{1 - \sigma}\right) \qquad (17)$$

Other nonselective examples are given, e.g., by Ross and Stonehart [93], by Larson and Hall [94], and by Ray et al. [95].

A recent example of reaction selectivity upon poisoning was reported by Dalla Betta et al. [96]: during heterogeneous methanation over Ru, Ni, and Rh, the introduction of H_2S into the system not only reduced the overall activity but also poisoned the ability of the sur-

SITE STRENGTH DISTRIBUTION

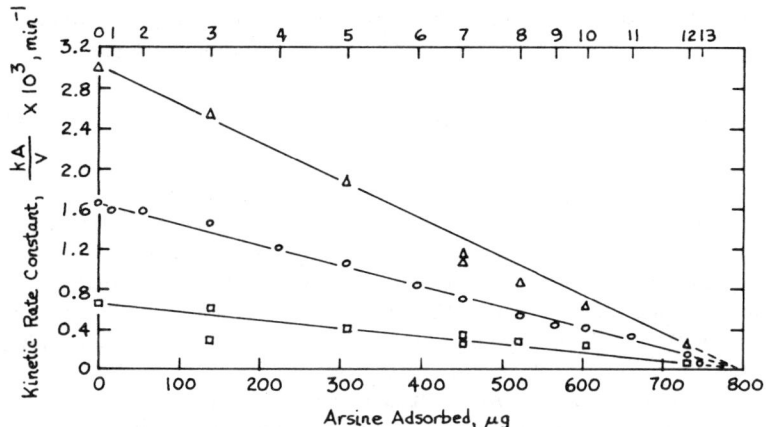

FIG. 21. Rate constants of cyclopropane hydrogenation over a Pt film as a function of the amount of AsH_3 adsorbed [92].

face to hydrogenate the carbon atom much more severely than the ability to form carbon-carbon bonds, thus significantly shifting the product distribution to higher molecular weight hydrocarbons.

Inoue et al. [97] studied the role of preadsorbed acetylene on the formation of benzene and the hydrogenation of ethylene and acetylene on polycrystalline Pd. Acetylene, adsorbed at 195 K, was found to desorb only partially (as benzene) from the Pd surface. Field emission microscopy showed that the partially poisoned surface contained vacant sites at stepped areas around the (111) facet which were active for hydrogenation of ethylene and acetylene. Furthermore, the authors found that the surface "template" formed by the adsorbed acetylene prevented ethylene decomposition and stabilized the activity of the catalyst for ethylene and acetylene hydrogenation.

The authors refer to a similar example where preadsorbed acetylene over Ni promoted the selectivity of cyclopropane isomerization to propylene, as opposed to decomposition into ethylene and ethane, or disproportionation into propane. Thus acetylene surface templates seem to offer reaction selectivity and activity stabilization on these polycrystalline surfaces.

For nonideal surfaces, such as those with site strength distribution and displaying selective or antiselective poisoning behavior, the highly nonlinear relationship between fraction of surface poisoned and fractional activity retained makes the development of detailed poisoning kinetic models very difficult. Szepe and Levenspiel [98] pointed out the advantages of a separable rate expression in the form

$$r_A(T, c, t) = r_1(T) r_2(c) r_3(a, t) \tag{18}$$

where $r_3(a, t)$ is a function of the remaining unpoisoned surface area a. Butt et al. [99] discussed the applicability of separable kinetics for catalysts of various nonuniform site strength distributions and concluded that nonideal surfaces cannot be described by a separable model, even if the site strength distributions are known. They found no correlation between the predictions of separable and nonseparable models for a variety of site strength distributions. Nevertheless, the assumption of separability often allows one to formulate simple expressions to correlate data for modeling purposes, and so the concept is of considerable engineering utility.

D. Multifunctional Catalysts

A large class of catalytic reaction networks requires the presence of multiple functions in the catalyst. Problems discussed in the previous part of this paper involved sites which had the same chemical nature but different strengths; we distinguish multicomponent catalysts by their property of carrying sites which differ in their chemical nature. Since each of the different functions may also possess its own site strength distribution, it is easy to see that the deactivation of multifunctional catalysts can be quite complicated.

There are two ways to induce multifunctionality. Metal oxide catalysts, for example, may inherently possess sites of differing chemical nature (acidic, basic). On the other hand, multifunctionality is also often accomplished by impregnating different metals onto the same support (e.g., automobile exhaust catalysts containing Pt, Pd, and Rh). The different metals may form alloys, thus further complicating the problem.

During the poisoning of multifunctional catalysts, the poison may preferentially deactivate certain sites; this phenomenon is referred to as <u>selective poisoning</u>. If the poison deactivates all sites, the phenomenon is termed <u>nonselective poisoning</u>.

The nonselective poisoning of bifunctional catalyst systems was mathematically analyzed by Butt [100], both for impurity poisoning and self-poisoning. The optimum formulation of the catalyst (i.e., the concentrations of the two individual functions) was shown to be different for best initial performance or for best long-time performance when deactivation occurs, and was shown to be a complex function of the activation energies of the main and deactivating reactions.

The poisoning of oxide catalysts containing multiple functions was recently reviewed by Knozinger [7]. Extremely interesting poisoning phenomena can occur here, since the chemical nature of the various sites is so different that most poisons show a strong selectivity toward one or the other. This phenomenon was exploited by a large number of workers to use specific poisoning experiments to determine the nature of the sites responsible for the different functions of the catalyst. Probably due to the industrial importance of these catalysts, the literature of their selective poisoning is so large that we can only refer to selected examples here, with emphasis on work subsequent to Knozinger's review [7].

γ-Alumina is frequently employed as a multifunctional catalyst, displaying both basic and acid character on different sites. Dabrowski et al. [101] employed a Monte Carlo technique to simulate the activity and selectivity properties of γ-alumina for dehydration; their work and references therein will serve to demonstrate the complexities involved with such multifunctional surfaces.

Various surface functional groups have been identified over aluminas, including lattice-terminating OH radicals and Al^{3+} ions. Adsorption studies have revealed Lewis and Bronsted acid sites and electron transfer sites. Specific poisoning experiments have contributed greatly to the understanding of which of these sites contribute to particular chemical reactions.

As an example, Saunders and Hightower [102] investigated deuterium exchange with benzene over γ- and η-aluminas. NO, CO, H_2O, O_2 NH_3, and CO_2 were explored as poisons. The strongest poison was CO_2 leading to the suggestion that H_2-D_2 equilibration, D_2 exchange, and H-D intermolecular redistribution all occur on Al^{3+} ions with a surface concentration of about 8×10^{12} cm^{-2}.

Four different sites were found over γ-alumina during deuterium exchange and isomerization experiments with olefins by Rosynek and Hightower [103], using a combination of selective CO_2 poisoning and temperature-programmed desorption experiments (Fig. 22).

Sites A and B chemisorbed unsaturated hydrocarbons so strongly that they did not undergo rapid communication with the gas phase. Sites A and B did not communicate appreciably, and neither was affected by CO_2. Site B equilibrated rapidly with surface OD groups, while Site A did so only to a limited extent.

The other two sites were associated with the isomerization activity (I sites) and exchange activity (E sites) of the catalyst. The

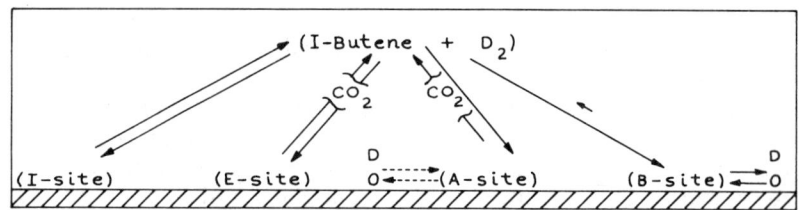

FIG. 22. Sites for deuterium exchange and isomerization on γ-alumina [103].

isomerization sites were not affected by CO_2 poisoning, while the exchange sites were effectively blocked, and the two sites were independent of each other, i.e., exchange occurred with isomerization.

Chuang et al. [104] investigated the CO_2 poisoning of the reaction

$$2COS + SO_2 \rightleftharpoons (3/x)S_x + 2CO_2 \tag{19}$$

over γ-alumina. Infrared spectroscopic experiments revealed that SO_2 is adsorbed on OH sites and COS is adsorbed on aluminum ion sites. The resulting mechanism is shown in Fig. 23. The aluminum ion sites were selectively poisoned by the irreversible chemisorption of CO_2, as shown in Fig. 24. Thus we have an interesting example of the site-selective self-poisoning of a multifunctional catalyst by one of the reaction products (CO_2).

The isomerization of 1-butene over amorphous alumina was studied by Ghorbel et al. [105]. Selective poisoning experiments revealed a very interesting complexity. For acid sites, NH_3 was used as the poison; for basic sites, acetic acid; for electron-accepting sites, phenothiazine; and for electron-donating sites, tetracyanoethylene. Any one of these poisons was effective in poisoning the isomerization activity of the catalyst, and thus it was concluded that acid sites of simultaneously oxidative nature and basic sites of simultaneously reducing nature are both required to catalyze the isomerization reaction.

The same reaction (i.e., isomerization of linear butenes) was investigated over silica (with Al impurities in the Al/Si atomic ratio range of 10^{-2} to 10^{-6}) by Van Roosmalen et al. [106]. Over such surfaces, surface Al impurities provide for Lewis acid sites and surface OH groups for Bronsted acid sites. Hexamethyldisilazane (a silylating agent) was employed to selectively poison the Bronsted acid sites,

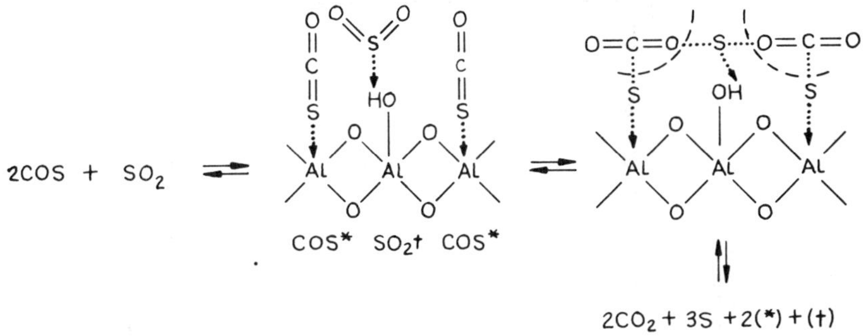

FIG. 23. CO_2 poisoning of the reaction $COS + SO_2$ [104].

which resulted in a strong suppression of the isomerization activity. From their selective poisoning experiments, the authors concluded that the isomerization activity is related to hydrogen-bonded silanol pairs (Bronsted acid sites) which interact with butene chemisorbed on the Lewis-acidic surface-aluminum ions.

The utility of selective poisoning experiments to find the sites responsible for various functions has been demonstrated by several recent papers. Aluminas were investigated by Knozinger et al. [107], Mochida et al. [108], Rosynek and Strey [109], and Hariharakrishnan et al. [110]; silica-aluminas by Take et al. [111], Mizuno et al. [112], Bakshi and Gavalas [113, 114], Lozos and Hoffman [115], and by Bremer et al. [116].

Chromia and lanthana catalysts were studied this way by Khodakov et al. [117], CaO and BaO by Hattori and Satoh [118] and Hattori et al. [119], Ga_2O_3 by Gilmore and Rooney [120], and chromia by Eley et al. [121], resulting in considerable insight into the complex nature of these processes.

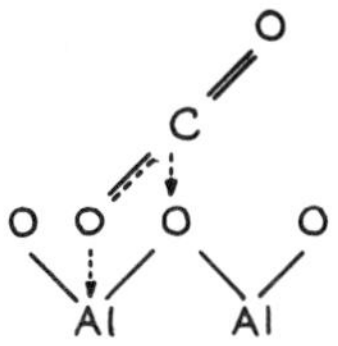

FIG. 24. CO_2 adsorption on alumina [104].

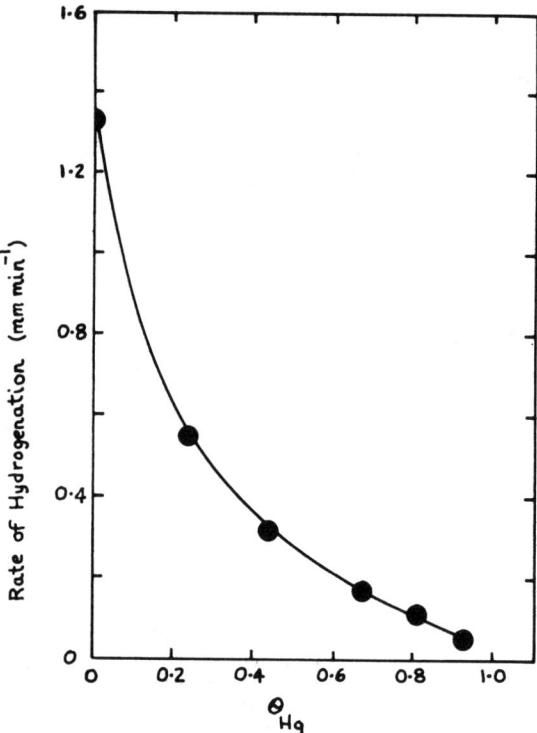

FIG. 25. 1-Butene hydrogenation rate as a function of Hg coverage over Rh [122].

Multifunctional catalysts are expected to show distinctly nonseparable kinetic behavior, and this was demonstrated by Bakshi and Gavalas [113]. For methanol and ethanol dehydration over fresh and n-butylamine-poisoned silica-aluminas, the rate data could be correlated by

$$r = \frac{kKc_A^{1/2}}{1 + K_A c_A^{1/2} + K_W c_W} \qquad (20)$$

where W refers to water. However, the numerical values of the parameters depended on the degree of deactivation and thus indicated

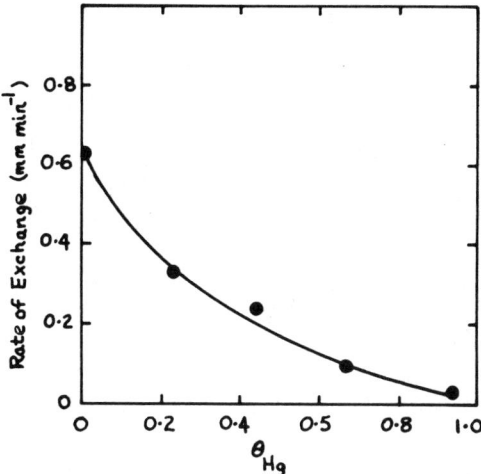

FIG. 26. 1-Butene exchange rate as a function of Hg coverage over Rh [122].

the presence of complexities (in this case, acid and base sites) which do not lend themselves to a description by separable poisoning kinetics. Very interesting effects may occur during the poisoning of such systems, such as the change of product distribution with flow reversal across a partially poisoned reactor.

Several supported metal catalysts display multifunctional character where different functions are carried by the metal and the support. The poisoning characteristics of these systems may again be complicated by the different sensitivity of the metal and the support toward poisoning; on the other hand, these complex poisoning characteristics allow both the selective manipulation of the individual functions and the elucidation of the reaction mechanism by intentional selective poisoning.

A large class of such problems is represented by noble metal-alumina catalysts used in various hydrocarbon conversion reactions.

Webb and Macnab [122] employed Hg poisoning, deuterium exchange, and radioactive tracer techniques to investigate the sites responsible for the hydrogenation, isomerization, and olefin exchange reactions of 1-butene over Rh. The rates of hydrogenation (Fig. 25) and deuterium exchange (Fig. 26) decreased gradually with increasing Hg coverage, while the rate of the isomerization reaction showed

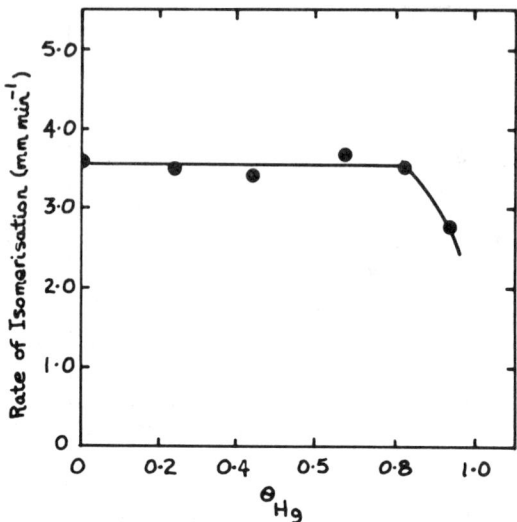

FIG. 27. 1-Butene isomerization rate as a function of Hg coverage over Rh [122].

no change with Hg coverage up to θ_{Hg} = 0.8 but declined thereafter (Fig. 27). The results were interpreted by a model which postulates that hydrogenation and olefin exchange occur directly on the metal, while isomerization involves the migration of adsorbed 1-butene from the metal to the silica support where the isomerization occurs.

The disproportionation of alkanes over a mixture of Pt-alumina and WO_3-silica was studied by Burnett and Hughes [123]. If excess olefin was used, the Pt component was completely deactivated (by coking), but the WO_3 component was left fully active.

Very interesting selective poisoning phenomena occur in catalytic reforming reactions. Sterba and Haensel [124] give an overview of historical developments in this area. Their examples include arsenic poisoning of Pt and its prevention by pretreating the feed, the modification of selectivity by sulfur poisoning, the use of selective poisoning of the various functions to elucidate the mechanism of the complex selectivity network at hand, and the use of bimetallic catalysts to combat deactivation by carbon formation.

Menon and Prasad [125] compared the sulfur poisoning (by thiophene) of Pt-alumina with that of Pt-Re-alumina during reforming reactions. The Pt in both catalysts converts to a sulfided state.

(The presulfidation of these catalysts is common industrial practice.) The sulfided catalysts are still active for dehydrogenation, dehydroisomerization, dehydrocyclization, and hydrocracking, but the relative contributions of these pathways are altered. Thus sulfidation decreases excessive dehydrogenation and so reduces coke formation. It also results in lowering the hydrocracking activity, leading to better selectivity toward aromatization and increased liquid yields.

In addition to the irreversibly bound sulfur, reversibly adsorbed sulfur may deposit at higher sulfur levels in the feed, resulting in a significant decrease in reforming activity. Thus low sulfur levels result in a beneficial selective poisoning while higher sulfur levels result in an undesirable poisoning, necessitating a control of sulfur concentration in the feedstream. The differences in sulfur tolerance between Pt-alumina and Pt-Re-alumina were interpreted in terms of the different relative contributions of reversibly and irreversibly held sulfur over those catalysts.

Pt-alumina and Pt-Ir-alumina reforming catalysts were compared by Ramaswamy et al. [126]. The "dilution" of Pt by an element with a lower dehydrogenation activity (Ir) leads to lower surface concentrations of coke precursors and hence to reduced coking rates. Presulfidation is required to reduce the increased hydrogenolysis activity of the Ir-containing catalyst, they conclude.

The literature of supported multifunctional catalysts is rich in interesting selective poisoning studies. Beyond the examples quoted above, let us mention here the work of Olsthoorn and Boelhouwer [127] on the Re_2O_7-alumina metathesis catalyst, the analysis of Co- and Mo-alumina systems by Lajacona et al. [128] and by Lee and Butt [129], and of Mo-alumina systems by Howe and Kemball [130], Balois and Beaufils [131], Cowley and Massoth [132], Millman and Hall [133], and Ratnasamy et al. [134]. Significant recent emphasis was paid to the elucidation of the mechanism of sulfur poisoning of various prospective multicomponent methanation catalysts; examples are the work of Shalvoy and Reucroft [135] on Ni-Cr-$MgSiO_3$ and of Wentrcek et al. [136] on Ni-Ir-alumina. In the latter example the authors showed by Auger electron spectroscopy that considerable Ir enrichment occurs on the surface of the Ni-Ir-alumina catalyst, exhibiting higher methanation activity and greater resistance to sulfur deactivation than Ni-alumina or Ru-alumina which were investigated for comparison. The effect of Ir in modifying the surface properties of Ni-alumina is well apparent from observing the temperature-programmed CO desorption spectra (Fig. 28): the addition of Ir to the Ni-Al_2O_3 catalyst resulted in more CO chemisorption and a higher-energy binding state in the presence of sulfur than observed for Ni-Al_2O_3 alone.

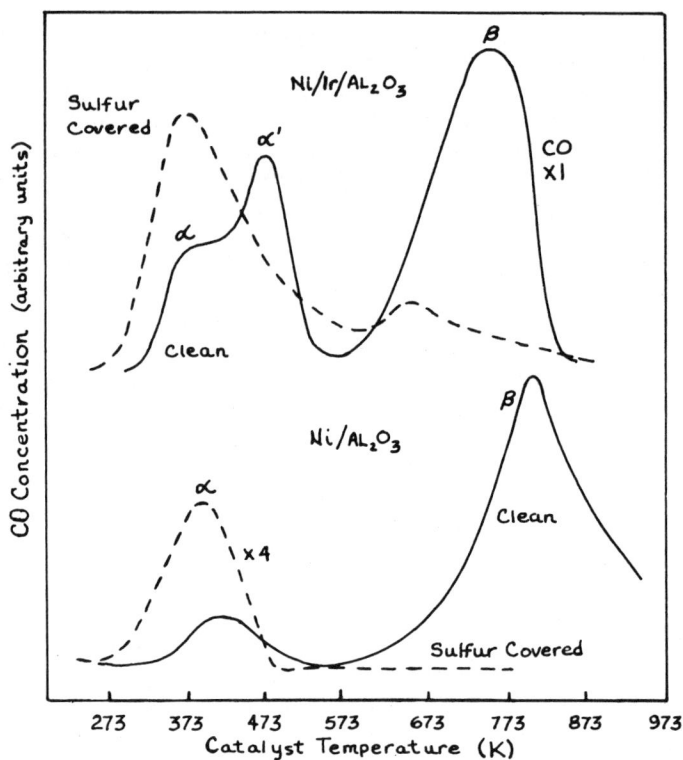

FIG. 28. Effect of Ir on the susceptibility of Ni toward the sulfur poisoning of CO adsorption (thermal desorption spectra) [136].

Several different reactions have to be simultaneously catalyzed in automobile exhaust converters, and these different reactions require the application of several catalytic components. These catalytic components have widely different poisoning characteristics and thus become poisoned to different extents. One of the more interesting features of these complex catalytic systems is the differing sulfur sensitivity of Pt, Pd, and Rh, resulting not only in different activity levels but also in differing selectivities toward the undesirable formation of NH_3. Thus, in a sulfur-free feedstream Pt produces large quantities of NH_3, which is selectively suppressed by the SO_2-content of the exhaust (Gandhi et al. [137], Summers and Baron [138]). Some other features of these complex multifunctional catalysts will be discussed in subsequent sections, and the review of Shelef [50] lists many more references.

To close our discussion of the poisoning of multifunctional catalysts, let us mention the brief review of Kemball [139], with some interesting examples of selective poisoning studies.

E. Support Effects

Various effects of catalyst supports on poisoning are discussed in different parts of this review; the space in this part is reserved to phenomena where the support does not participate directly in the main reactions (i.e., it does not possess sites which promote them), but participates in or influences in some way the poisoning process.

Even without poisoning, the interactions between supports and catalytic components are very complicated and represent much of the current interest in the frontiers of catalytic research. Needless to say, poisoning considerations add further complexities which are often only partially understood. Nevertheless it seems to be worthwhile to discuss some examples, even though a rigid classification may be somewhat premature at this stage.

The support may often adsorb a poison or react with it. Typical examples can be taken from automobile exhaust catalysis, where Pb-, P-, and S-containing impurities in the feedstream are quite reactive with the γ-alumina support commonly employed in this application. A complex variety of species may form, such a lead aluminate, aluminum-o-phosphate, lead sulfate, and chemisorbed SO_2 and SO_3. In transport-influenced systems such as these, the rate of deactivation of the catalyst pellet can be manipulated by employing the "poison-getter" properties of the support to protect the catalytically active components; such examples will be discussed in detail in the subsequent sections.

Upon reacting with the support, the poisons may significantly alter its characteristics. Mooi et al. [140] provide an example: at high temperatures, lead in automobile exhaust interacts with the cordierite (Mg-Al-silicate) substrate of catalytic monoliths, resulting in enhancing their rate of thermal decomposition and thus lowering their melting point. Mullite, γ-alumina, and $PbAlSiO_4$ have been detected in such systems at high temperatures.

Dalla Betta et al. [96] investigated the H_2S poisoning of the methanation of CO over Ru, Ni, and Rh. In comparison with ZrO_2- and Al_2O_3-supported Ni, the unsupported Raney Ni catalyst was severely deactivated. Since Raney Ni contains some Al, the authors suggested that the formation of $NiAl_2S_4$ (sulfospinels) on the surface of the Raney Ni catalyst may explain this large difference between the supported and unsupported Ni.

The methanation catalysts $ThNi_5$ and $ZrNi_5$ were the subject of the poisoning experiments of Moldovan et al. [141]. In use, $ZrNi_5$ transformed into Ni supported on ZrO_2, and $ThNi_5$ into Ni supported on ThO_2. The ThO_2-supported catalyst was resistant to self-poisoning while the ZrO_2-supported catalyst was heavily covered with graphite. The authors concluded that Ni/ThO_2 is sufficiently active to convert adsorbed carbon complexes into hydrocarbons while Ni/ZrO_2 is not.

$ThNi_5$ was poisoned by H_2S which converted the active Ni to Ni_3S_2. $ZrNi_5$ was more sensitive to sulfur poisoning; this was attributed to its lower surface-Ni concentration.

Sometimes small quantities of additives may result in a drastic alteration of poison resistance. This was shown, for example, by Johnson et al. [142] and Gallagher et al. [143]: when Pt traces were added to $La_{0.7}Pb_{0.3}MnO_3$, its resistance to SO_2 poisoning (during C_2H_6 and CO oxidation) was enhanced. In fact, the addition of SO_2 enhanced the activity of one of the catalysts: for an explanation the authors quote the observations of Yao [144] who found that SO_2 effectively converts a less active PtPb alloy (which may have formed here) to a more active combination of Pt and segregated $PbSO_4$.

Supports may interact with the catalytic metals in such a way that sites with different strengths are created; in turn, due to these different sites, supported catalysts may exhibit a poisoning behavior which does not occur on unsupported catalysts. An example for this phenomenon is the sulfur poisoning of Ni.

On single crystal Ni(111) surfaces, Erley and Wagner [23] found complete sulfur poisoning of CO chemisorption above a sulfur coverage of 0.3. On the other hand, Rochester and Terrell [29], investigating $Ni-SiO_2$, and Rewick and Wise [45], investigating $Ni-Al_2O_3$, found that CO adsorption is not completely inhibited by H_2S. Their infrared spectra over sulfur-free and sulfur-covered surfaces reveal that sites ascribed to more than one CO bonded to one Ni atom (plausibly associated with Ni atoms of low coordination) may be responsible for this apparent increase in sulfur resistance. Since all Ni atoms are highly coordinated in Ni(111), the conclusion is that the SiO_2 and Al_2O_3 supports created low-coordination Ni sites and thus favorably affected the poison resistance of these catalysts.

Another example of structure-sensitive poisoning studies is the work of Barbier et al. [145]. A series of Pt-alumina catalysts were prepared with dispersions ranging from 5% to 80%. Benzene hydrogenation, cis-1,2-dimethyl cyclohexane epimerization, cyclopentane hydrogenolysis, benzene exchange, cyclopentane monoexchange, and cyclopentane multiexchange were employed as test reactions, and a wide variety of reversible and irreversible poisons were explored to probe the structure sensitivity of the poisoning of these reactions.

Benzene hydrogenation over the unpoisoned catalysts was found to be structure insensitive. However, its inhibition by NH_3 caused

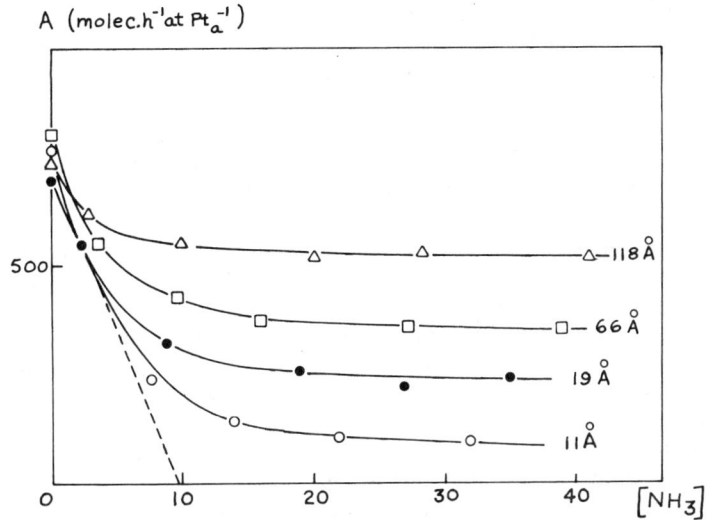

FIG. 29. Effect of Pt particle size on the NH_3 inhibition of benzene hydrogenation [145].

a strong structure sensitivity (Fig. 29): for large crystallites the turnover numbers were less sensitive to NH_3 inhibition than for small crystallites. Thus the structure sensitivity of the poison adsorption imparted an apparent structure sensitivity to the main reaction in the presence of a poison in the feedstream. Manogue and Katzer [146] call such phenomena "secondary structure sensitivity" as distinguished from the "primary structure sensitivity" which is reserved for the structure sensitivity of the main reactions in the absence of poisons. Cyclopentane hydrogenolysis would serve as an example in Barbier et al. [145] for primary structure sensitivity.

The degree of metal dispersion may often influence the poison resistance of catalysts, even if structure sensitivity arguments are not necessarily involved. For example, Boudart et al. [147], comparing the specific activities of Pt-alumina catalysts across a wide range of dispersions for the hydrogenation of cyclopropane, found a marked susceptibility of the well-dispersed catalysts toward oxygen poisoning, even though the poison-free specific reaction rates varied only by a modest factor of 2 across the range of dispersions. They hypothesized that water, formed over the Pt, stays on the support in the immediate vicinity of a Pt site but reduces the activity by some sort of interaction. Interestingly enough, Iida and Tamaru [148] report that the H_2O poisoning of the D_2-H_2O exchange reaction over Pt can be significantly reduced if a hydrophobic support (Porapak Q) is employed.

Ethylene hydrogenation was studied by Dorling et al. [149] over Pt-SiO_2, varying the Pt crystallite size from 13 to 200 Å. Again, small Pt areas were found to be easily poisoned, in their case by self-poisoning due to carbonaceous residues.

Hegedus and Petersen [150], who investigated the self-poisoning of cyclopropane hydrogenation over Pt-Al_2O_3, made similar observations regarding particle size effects. For larger crystallites the catalyst could not be completely poisoned. The data could be intrepreted by the assumption of a ring of carbonaceous residues on the Pt islands, surrounding their still active portion. If this ring can only grow to a finite thickness, islands larger than a certain size cannot be completely covered by these carbonaceous residues.

The reactants may communicate between the metal and the support, resulting in a phenomenon termed "spillover." Thus the diffusion of hydrogen from Pt to a carbon surface can be facilitated by carbonaceous contaminants: once they are removed, the diffusion of hydrogen atoms to the carbon surface stops (Boudart et al. [151]). Such phenomena may play an important role in the self-poisoning of supported noble metal catalysts due to partially dehydrogenated carbonaceous residues, even though some counterexamples may exist (Fuentes and Figueras [152]).

An interesting particle size-dependent self-poisoning phenomenon was reported by Ostermaier et al. [153], who investigated the oxidation of NH_3 over Pt. The size of Pt particles covered the range of 20 to 150 Å. During reaction conditions, the catalyst was deactivated by oxidation of the Pt crystallites to an extent of several atomic layers in depth. Small Pt crystallites were almost completely deactivated, while larger ones were more resistant. The proximity of noble metals to the support (i.e., their dispersion) may result in enhancing their oxidizability (e.g., Joyner [154]).

There are several additional interesting support effects which deserve mention here. Thus Gallezot et al. [155] reported that Pt-Y zeolites showed a remarkable resistivity toward sulfur poisoning and a sensitivity toward NH_3 poisoning during benzene hydrogenation, which was attributed to an electron-deficient character of Pt in such structures.

3
Intra- and Interparticle Transport Effects

A. Introductory Remarks

Transport effects have a profound influence on the performance and durability of catalysts. Since their consideration is both of scientific and industrial importance, the careful analysis of these effects is an essential part of catalytic research.

Surface transport during catalyst poisoning (solid-state diffusion and surface diffusion) was discussed in the previous section, so we will concentrate here on fluid-phase transport effects. Of these, we will restrict our attention to the transport of reactants and poison precursors on the scale of a single catalyst pellet, while referring to the reviews of Butt [5], Butt and Billimoria [8], and Butt [10] on the effects of transport resistances on the poisoning of catalytic reactors.

B. Mechanistic Considerations

The first systematic analysis of the poisoning of diffusion-influenced catalyst pellets is due to Wheeler [156]. In his often-quoted and now classical analysis of the shape of catalyst poisoning curves (relationship between fractional activity and fraction of surface poisoned), he demonstrated that a nonlinear relationship between catalyst activity and fraction of surface poisoned can be effectively mimicked by intra-pellet diffusion resistances, even for the case where all original sites are equivalent and single-site mechanisms prevail.

Wheeler's analysis has an interesting historical connotation. The "active site" concept of catalysis (e.g., Taylor [87]), that is, the suggestion that only a certain fraction of the catalyst's surface is active, was largely motivated by the observation (e.g., Constable [88]) that often very small quantities of the poison can nearly completely deactivate a catalyst, corresponding to only a small fraction of the total available surface being covered by the poison. The implication was that the active sites preferentially react with the poison precursor. While in many cases this is indeed true, the same apparent phenomenon can arise due to intrapellet diffusion limitations.

Intraparticle diffusion effects introduce concentration gradients in the catalyst pellets, while purely kinetic control implies uniform concentration profiles. According to the main controlling resistance, the simple situations of Table 2 can be considered. Let us discuss these individual cases, following Wheeler's reasoning, with appropriate extensions.

TABLE 2

	Main reaction	Poisoning reaction
A	Kinetic	Kinetic
B	Diffusion	Kinetic
C	Kinetic	Diffusion
D	Diffusion	Diffusion (pore mouth poisoning)
E	Diffusion	Diffusion (core poisoning)

For the sake of simplicity, a first-order, irreversible, monomolecular, isothermal reaction is taken to occur in a flat, one-dimensional, porous catalyst slab. All catalytic sites are equivalent and, at a given time, a fraction α of them has been irreversibly poisoned. The reaction rate per catalyst pellet is $\widetilde{R}$ (mol/s), the reaction rate for the poison-free catalyst pellet is $\widetilde{R}_o$, and thus $\widetilde{R}/\widetilde{R}_o$ represents the fractional activity remaining.

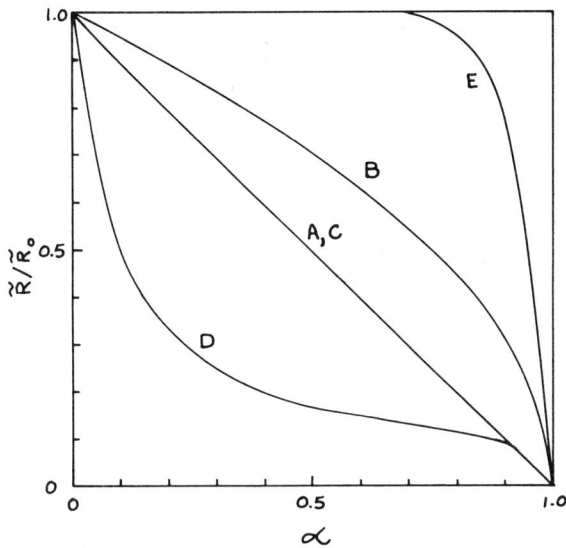

FIG. 30. Relative reaction rate as a function of the fraction of catalyst poisoned. See text for explanation.

In Case A, both the reactant and the poison are uniformly distributed in the pellet, so

$$\tilde{R}/\tilde{R}_o = 1 - \alpha \tag{21}$$

a straight diagonal line in Fig. 30 (Curve A), termed nonselective poisoning.

In Case B, the poison is uniformly distributed in the pellet, but finite diffusion effects on the main reaction cause intrapellet concentration gradients. Solution of the appropriate diffusion-reaction equation yields

$$\frac{\tilde{R}}{\tilde{R}_o} = \frac{\sqrt{1-\alpha}\, \tanh(h_o \sqrt{1-\alpha})}{\tanh h_o} \tag{22}$$

where h_o is the Thiele parameter of the poison-free pellet. At the limit of complete intrapellet diffusion control,

$$\lim_{h_o \to \infty} (\tilde{R}/\tilde{R}_o) = \sqrt{1-\alpha} \tag{23}$$

mimicking an "antiselective" poisoning behavior (Fig. 30, Curve B). At the limit of a kinetically controlled main reaction,

$$\lim_{h_o \to 0} (\widetilde{R}/\widetilde{R}_o) = 1 - \alpha \tag{24}$$

the straight diagonal line in Fig. 30.

Case C turns out to be similar to Case A: since the main reaction is kinetically controlled, the relationship between relative rate and fraction poisoned is simply

$$\widetilde{R}/\widetilde{R}_o = 1 - \alpha \tag{25}$$

our straight diagonal line (Fig. 30, Curve C). Thus, if the main reaction is kinetically controlled, the reaction rate declines linearly with increasing α no matter how the poison is distributed in the pellet.

Case D represents pore-mouth poisoning where the outer portion of the catalyst pellet has been poisoned by a sharply defined poison front. Solution of the appropriate diffusion-reaction problem yields

$$\frac{\widetilde{R}}{\widetilde{R}_o} = \frac{\tanh[h_o(1-\alpha)]}{\tanh h_o} \left(\frac{1}{1 + \alpha h_o \tanh[h_o(1-\alpha)]}\right) \tag{26}$$

This function converges to $\widetilde{R}/\widetilde{R}_o = 0$ as $h_o \to \infty$. At large values of h_o, a strongly "selective" poisoning curve is mimicked by pore-mouth poisoning, as shown by Curve D in Fig. 30, computed for $h_0 = 10$, a moderately strong intrapellet diffusion resistance.

At the limit of a kinetically controlled main reaction,

$$\lim_{h_o \to 0} (\widetilde{R}/\widetilde{R}_o) = 1 - \alpha \tag{27}$$

corresponding to Case C discussed previously. On the other hand, at the limit of complete diffusion control,

$$\lim_{h_o \to \infty} (\widetilde{R}/\widetilde{R}_o) = 0 \tag{28}$$

implying the increasing severity of pore-mouth poisoning as the degree of diffusion control increases.

MECHANISTIC CONSIDERATIONS 51

Case E describes a situation where the central portion of the pellet has been poisoned by a sharply defined poison front (core poisoning). Such situations may arise in diffusion-influenced self-poisoning reactions where the reaction product acts as the poison precursor. Solution of the appropriate diffusion-reaction problem yields

$$\frac{\tilde{R}}{\tilde{R}_0} = \frac{\tanh[h_0(1-\alpha)]}{\tanh h_0} \qquad (29)$$

which is displayed for $h_0 = 10$ as Curve E in Fig. 30, mimicking a strong antiselective poisoning behavior.

At the limit of complete diffusion control,

$$\lim_{h_0 \to \infty} (\tilde{R}/\tilde{R}_0) = 1 \qquad (30)$$

implying that the severity of core poisoning declines with increasing intrapellet diffusion resistances. On the other hand, for a kinetically controlled main reaction, an application of l'Hopital's rule yields

$$\lim_{h_0 \to 0} (\tilde{R}/\tilde{R}_0) = 1 - \alpha \qquad (31)$$

the diagonal nonselective line in Fig. 30.

Thus we can conclude that the shape of the $\tilde{R}/\tilde{R}_0$-α curves (Fig. 30) contains some important mechanistic information which, if α can be reliably determined, can serve as a diagnostic tool. However, the fraction of catalyst pellet surface poisoned may not always be easy to evaluate experimentally. An alternate technique has been developed by Petersen and collaborators (reviewed by Hegedus and Petersen [157] and Petersen [158]) involving a single-pellet diffusion reactor.

Realizing that the concentration of reactants at the center of a catalyst pellet differs from their concentration at the pellet's outer surface if intrapellet diffusion resistances prevail, Balder and Petersen [159, 160] showed that center-plane concentration measurements can provide useful information on the degree of diffusion resistances in the pellet. A normalized center-plane concentration

$$\Phi_A = \frac{\psi_A(t, 1) - \psi_A(0, 1)}{1 - \psi_A(0, 1)} \qquad (32)$$

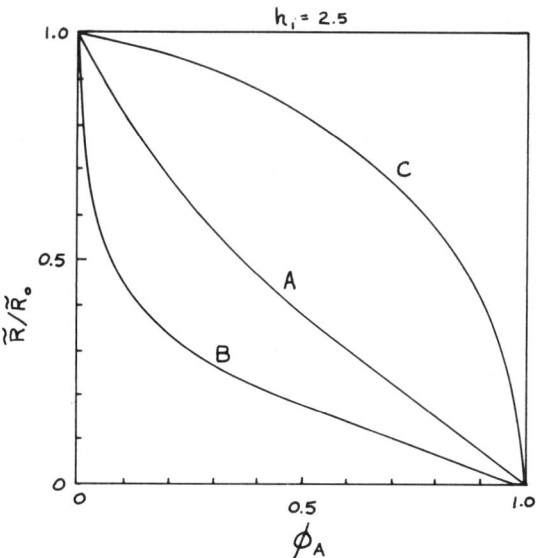

FIG. 31. Relative reaction rates as a function of normalized center-plane concentration in a single-pellet diffusion reactor. See text for explanation [162].

was introduced, where $\psi_A(1)$ is the center-plane concentration of reactant A divided by the concentration of A at the pellet's outer surface. $t = 0$ refers to the fresh catalyst pellet, and $t = t$ to the pellet poisoned to a certain extent. The idea is to follow $\tilde{R}/\tilde{R}_o$ as a function of Φ_A during the course of the poisoning process (see also Dougharty [161]).

The corresponding diffusion-reaction problem can be easily solved for various reaction orders and poisoning mechanisms. The resulting plot of $\tilde{R}/\tilde{R}_o$ vs Φ_A (Fig. 31) allows one to discriminate among various poisoning mechanisms. All curves in Fig. 31 were computed for a main reaction Thiele parameter (h_1) of 2.5, a convenient degree of intrapellet diffusion resistance which allows the accurate measurement of center-plane concentrations. Curve A is for uniform poisoning, Curve B is for pore-mouth poisoning, and Curve C is for core poisoning. For a given value of h_1, impurity poisoning and parallel self-poisoning fall somewhere between Curves A and B, series self-poisoning between Curves A and C, and triangular self-poisoning (i.e., simultaneous parallel and series self-poisoning) between Curves B

MECHANISTIC CONSIDERATIONS

and C, depending on the relative contributions of the parallel and series paths. Restricted regions apply to self-poisoning reactions within the above-described boundaries.

Various types of single-pellet diffusion reactors have been built. All involve a flat catalyst slab. One face is exposed to the bulk reactant stream while the other face is terminated by a closed space, the center-plane chamber. At steady-state operation the concentration in the center-plane chamber equals the concentration of the gas in the pellet's pores at the center of a catalyst slab twice as thick as the one installed in the reactor. The center-plane chamber can be monitored by gas chromatography (Balder and Petersen [159]), infrared spectroscopy (Hegedus and Petersen [163]), or by mass spectroscopy (Tennant and Wei [164] and Herz [165]). The reactor of Hegedus and Petersen [163] is shown in Fig. 32.

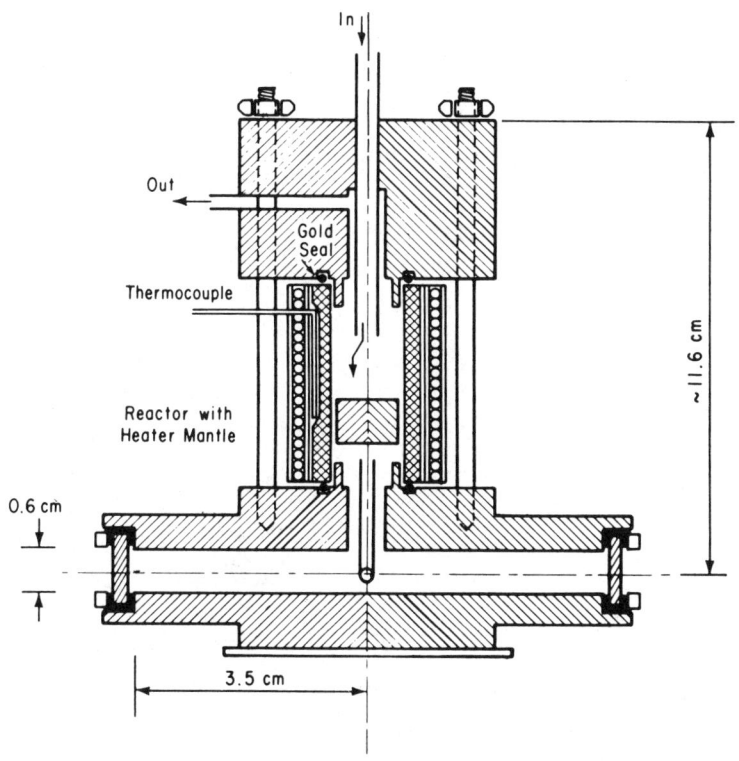

FIG. 32. Single-pellet diffusion reactor of Hegedus and Petersen [163].

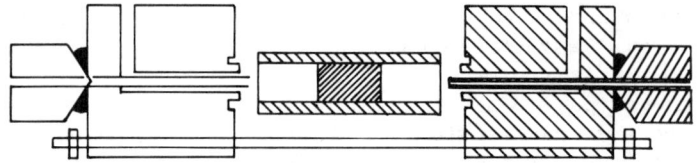

FIG. 33. Single-pellet diffusion reactor of Hahn and Petersen [166].

Nonuniform activity profiles in the pellet due to poisoning were demonstrated in a simple experiment by Hahn and Petersen [166]: different activities were observed when the exposed pellet face and the center-plane chamber were interchanged after some extent of diffusion-influenced poisoning (Figs. 33 and 34).

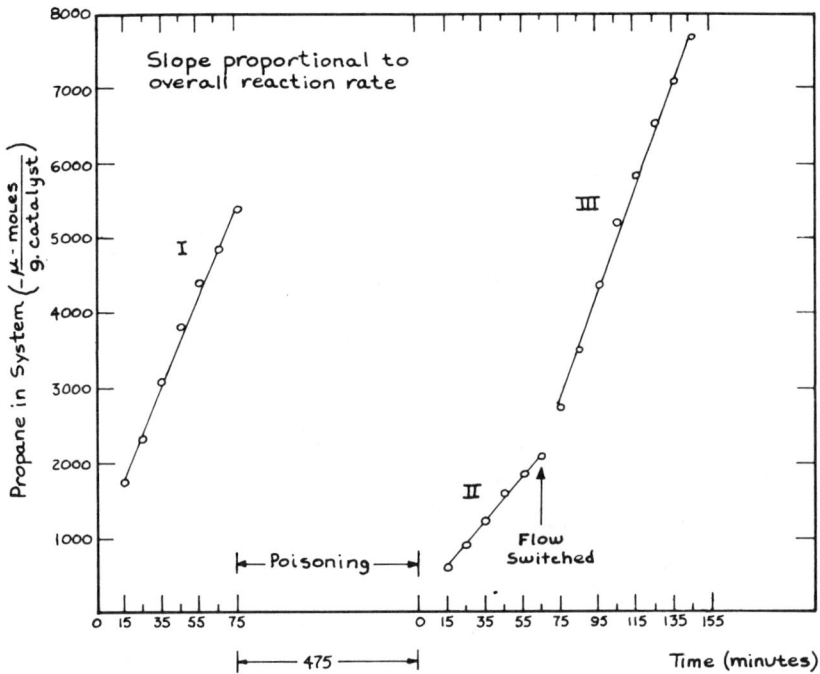

FIG. 34. Demonstration of nonuniform poisoning by switching the exposed pellet face [166].

MECHANISTIC CONSIDERATIONS

Hegedus and Petersen [150, 163, 167] investigated the deactivation of cyclopropane hydrogenation (to propane) over a Pt-alumina catalyst using a single-pellet diffusion reactor. The analysis showed that the catalyst is deactivated simultaneously by both the reactant and the product (triangular self-poisoning) according to the mechanism

(33)

which was quantified by

$$\frac{\partial^2 \psi_A}{\partial \eta^2} - h_1^2 \theta \psi_A = 0 \tag{34}$$

$$\theta^\alpha \psi_A + \frac{k_3}{k_2} \theta^\alpha (\omega - \psi_A)^\delta = -\frac{\partial \theta}{\partial \tau} \tag{35}$$

$$\psi_A(0, \eta) = \frac{\cosh [h_1(1 - \eta)]}{\cosh h_1} \tag{36}$$

$$\theta(0, \eta) = 1 \tag{37}$$

$$\psi_A(\tau, 0) = 1 \tag{38}$$

$$\frac{\partial \psi_A}{\partial \eta}(\tau, 1) = 0 \tag{39}$$

and

$$\omega = \psi_A + \psi_B = \text{constant} \tag{40}$$

The assumptions inherent in this analysis will be apparent later on. Figure 35 compares the predictions of this model (for $\alpha = 1$, $\delta = 2$, $k_3/k_2 = 0.4$, $h_1 = 3.809$) with experimental results, with the conclusion that the catalyst deactivates by a triangular self-poisoning mechanism.

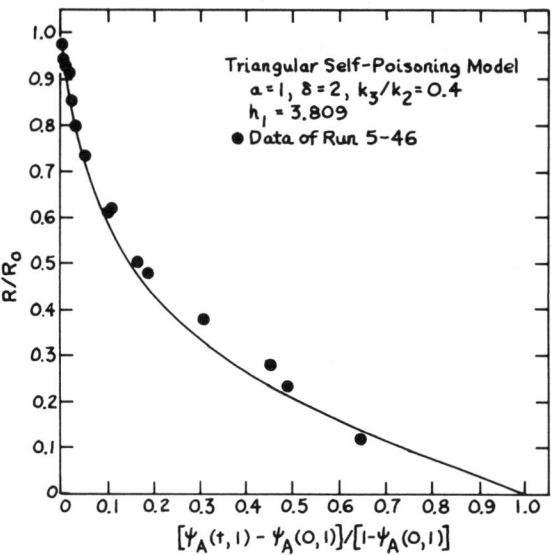

FIG. 35. Triangular self-poisoning during cyclopropane hydrogenation in a single-pellet diffusion reactor [167].

The single-pellet diffusion reactor was subsequently employed by Wolf and Petersen [168] to analyze the deactivation of methylcyclohexane dehydrogenation over Pt-alumina in the temperature range of 350-400°C, demonstrating the utility of the technique under more severe experimental conditions. The mechanistic and kinetic insight which can be gained from such experiments, coupled with the economy of experimentation resulting from the sensitivity of center-plane concentrations to deactivating events, make this technique especially valuable for catalyst poisoning studies. Plots such as Fig. 35 eliminate time as an explicit variable, and focus the attention on the mechanism of the deactivating event. In principle, main reactions of arbitrary order can be easily analyzed (Wolf and Petersen [169]); interesting singularities are introduced for a zero-order main reaction (Wolf and Petersen [170]).

C. Time-Dependent Behavior

Now that we gained a general impression about the effects of intrapellet diffusion resistances on the course of catalyst poisoning, let us venture into the corresponding time-dependent problem. We will first derive a somewhat general case and reduce it later on to specific formulations for particular problems.

To keep matters reasonably simple, let us consider a porous catalyst pellet in which the isothermal, monomolecular, first-order conversion of A is inhibited by an impurity P or poisoned by a species W which results from an irreversible, single-site reaction of chemisorbed P with a catalytic site. The resulting mechanism is

$$A + S \rightleftharpoons AS \tag{41}$$

$$AS \rightleftharpoons B + S \tag{42}$$

$$P + S \rightleftharpoons PS \tag{43}$$

$$PS \longrightarrow WS \tag{44}$$

where all sites S are equivalent and B does not chemisorb.

The transient diffusion-reaction problem is described by the following conservation equations:

$$D_A \nabla^2 c_A - a \left[\frac{\partial \theta_A}{\partial t} + k_A \theta_A \right] = \epsilon \frac{\partial c_A}{\partial t} \tag{45}$$

$$\partial \theta_A / \partial t = k_{a,A} (1 - \theta_A - \theta_P - \theta_W) c_A - k_{d,A} \theta_A - k_A \theta_A \tag{46}$$

$$D_P \nabla^2 c_P - a \left[\frac{\partial \theta_P}{\partial t} + k_P \theta_P \right] = \epsilon \frac{\partial c_P}{\partial t} \tag{47}$$

$$\partial \theta_P / \partial t = k_{a,P} (1 - \theta_A - \theta_P - \theta_W) c_P - k_{d,P} \theta_P - k_P \theta_P \tag{48}$$

$$\partial \theta_W / \partial t = k_P \theta_P \tag{49}$$

with appropriate initial and boundary conditions.

Note that we allowed A and P to compete for the free sites, but assumed that once WS has formed it irreversibly occupies a fraction θ_W of the sites. Thus our formulation incorporates both reversible inhibition and irreversible poisoning.

The equilibrium assumption is often invoked at this point to express θ_A and θ_P in terms of c_A and c_P. The validity of this assumption has to be carefully analyzed for individual cases but, for slow transients, it is frequently utilized. If so, Eqs. (45) to (49) will collapse into

$$D_A \nabla^2 c_A - a\left[\frac{\partial \theta_A}{\partial c_A}\frac{\partial c_A}{\partial t} + \frac{\partial \theta_A}{\partial c_P}\frac{\partial c_P}{\partial t} + \frac{\partial \theta_A}{\partial \theta_W}\frac{\partial \theta_W}{\partial t} + k_A \theta_A\right] = \epsilon \frac{\partial c_A}{\partial t} \quad (50)$$

$$D_P \nabla^2 c_P - a\left[\frac{\partial \theta_P}{\partial c_A}\frac{\partial c_A}{\partial t} + \frac{\partial \theta_P}{\partial c_P}\frac{\partial c_P}{\partial t} + \frac{\partial \theta_P}{\partial \theta_W}\frac{\partial \theta_W}{\partial t} + k_P \theta_P\right] = \epsilon \frac{\partial c_P}{\partial t} \quad (51)$$

$$\partial \theta_W / \partial t = k_P \theta_P \quad (52)$$

where the appropriate isotherms and their partial derivatives can be substituted. For our examples, we will use Langmuirian isotherms which, for the case of irreversibly poisoned θ_W, work out to

$$\theta_A = \frac{K_A c_A (1 - \theta_W)}{1 + K_A c_A + K_P c_P} \quad (53)$$

$$\theta_P = \frac{K_P c_P (1 - \theta_W)}{1 + K_A c_A + K_P c_P} \quad (54)$$

If the reversible inhibition by the poison precursor P is to be modeled, $k_P = \theta_W = 0$, and Eqs. (50) to (52) will simplify to

$$D_A \nabla^2 c_A - a\left[\frac{\partial \theta_A}{\partial c_A}\frac{\partial c_A}{\partial t} + \frac{\partial \theta_A}{\partial c_P}\frac{\partial c_P}{\partial t} + k_A \theta_A\right] = \epsilon \frac{\partial c_A}{\partial t} \quad (55)$$

$$D_P \nabla^2 c_P - a\left[\frac{\partial \theta_P}{\partial c_A}\frac{\partial c_A}{\partial t} + \frac{\partial \theta_P}{\partial c_P}\frac{\partial c_P}{\partial t}\right] = \epsilon \frac{\partial c_P}{\partial t} \quad (56)$$

Further simplifications are often possible. For example, the main reaction is often assumed to relax much faster than the inhibition problem, resulting in

$$D_A \nabla^2 c_A - ak_A \theta_A = 0 \quad (57)$$

TIME-DEPENDENT BEHAVIOR

$$D_P \nabla^2 c_P - a\left[\frac{\partial \theta_P}{\partial c_A}\frac{\partial c_A}{\partial t} + \frac{\partial \theta_P}{\partial c_P}\frac{\partial c_P}{\partial t}\right] = \epsilon\frac{\partial c_P}{\partial t} \qquad (58)$$

Further simplifying assumptions may neglect the influence of A on the adsorption of P, giving

$$\frac{\partial \theta_P}{\partial c_A}\frac{\partial c_A}{\partial t} = 0 \qquad (59)$$

and, if applicable, $K_A c_A \ll 1$, which together yield

$$D_A \nabla^2 c_A - \frac{ak_A K_A c_A}{1 + K_P c_P} = 0 \qquad (60)$$

$$D_P \nabla^2 c_P = \frac{\partial c_P}{\partial t}\left[\epsilon + \frac{aK_P}{(1 + K_P c_P)^2}\right] \qquad (61)$$

a formulation analogous to the problem discussed by Gioia et al. [171]. In addition,

$$\epsilon \ll \frac{aK_P}{(1 + K_P c_P)^2} \qquad (62)$$

also often seems appropriate; in that case,

$$D_A \nabla^2 c_A - \frac{ak_A K_A c_A}{1 + K_P c_P} = 0 \qquad (63)$$

$$D_P \nabla^2 c_P = \frac{\partial c_P}{\partial t}\left[\frac{aK_P}{(1 + K_P c_P)^2}\right] \qquad (64)$$

arises as a "minimum description" of the time-dependent inhibition problem.

For <u>irreversible poisoning</u>, k_P = finite, but the <u>quasi-steady-state assumption</u> (Bischoff [172]) can be invoked quite often; in this case it states that the relaxation of A and P are much faster than that of WS. In terms of Eqs. (45) to (49), this translates into

$$\frac{\partial \theta_A}{\partial c_A} \frac{\partial c_A}{\partial t} = \frac{\partial \theta_A}{\partial c_P} \frac{\partial c_P}{\partial t} = \epsilon \frac{\partial c_A}{\partial t} = 0 \tag{65}$$

$$\frac{\partial \theta_P}{\partial c_A} \frac{\partial c_A}{\partial t} = \frac{\partial \theta_P}{\partial c_P} \frac{\partial c_P}{\partial t} = \epsilon \frac{\partial c_P}{\partial t} = 0 \tag{66}$$

which yields

$$D_A \nabla^2 c_A - a \left[\frac{\partial \theta_A}{\partial \theta_W} \frac{\partial \theta_W}{\partial t} + k_A \theta_A \right] = 0 \tag{67}$$

$$D_P \nabla^2 c_P - a \left[\frac{\partial \theta_P}{\partial \theta_W} \frac{\partial \theta_W}{\partial t} + k_P \theta_P \right] = 0 \tag{68}$$

$$\partial \theta_W / \partial t = k_P \theta_P \tag{69}$$

Considering the isotherms of A and P (Eqs. 53 and 54), we first obtain

$$\frac{\partial \theta_A}{\partial \theta_W} = \frac{- K_A c_A}{1 + K_A c_A + K_P c_P} \tag{70}$$

and

$$\frac{\partial \theta_P}{\partial \theta_W} = \frac{-K_P c_P}{1 + K_A c_A + K_P c_P} \tag{71}$$

which, when substituted into (67) to (69), yield

$$D_A \nabla^2 c_A - \frac{ak_A K_A c_A (1 - \theta_W)}{1 + K_A c_A + K_P c_P} \left[1 - \frac{(k_P/k_A) K_P c_P}{1 + K_A c_A + K_P c_P} \right] = 0 \tag{72}$$

$$D_P \nabla^2 c_P - \frac{ak_P K_P c_P (1 - \theta_W)}{1 + K_A c_A + K_P c_P} \left[1 - \frac{K_P c_P}{1 + K_A c_A + K_P c_P} \right] = 0 \tag{73}$$

$$\frac{\partial \theta_W}{\partial t} = k_P \frac{K_P c_P (1 - \theta_W)}{1 + K_A c_A + K_P c_P} \tag{74}$$

For

$$K_A c_A \approx K_P c_P \ll 1 \tag{75}$$

and

$$(k_P/k_A) K_P c_P \ll 1 \tag{76}$$

the formulation reduces to

$$D_A \nabla^2 c_A - a k_A K_A c_A (1 - \theta_W) = 0 \tag{77}$$

$$D_P \nabla^2 c_P - a k_P K_P c_P (1 - \theta_W) = 0 \tag{78}$$

$$\partial \theta_W / \partial t = k_P K_P c_P (1 - \theta_W) \tag{79}$$

which is analogous to the classical analysis of impurity poisoning by Masamune and Smith [173].

The literature is rich in papers which investigate various time-dependent, isothermal, diffusion-influenced inhibition and poisoning problems. Many of these were reviewed by Butt [5, 8, 10]. Beyond the papers of Gioia et al. [171] and Masamune and Smith [173] mentioned above, let us quote here the elegant analyses of Carberry and Gorring [174] and Olson [175] of time-dependent pore-mouth poisoning, the work of Chu [176] which explored Langmuir-Hinshelwood models to triangular self-poisoning, Murakami et al. [177] who showed both experimentally and theoretically how a diffusion-influenced series self-poisoning mechanism leads to core poisoning, and the analysis of reversible inhibition by an impurity due to Gioia et al. [178] and Gioia [179]. A molecular view of diffusion-influenced poisoning was provided by Merrill [180]. Hegedus [181] and Zwicky and Gut [182] investigated impurity poisoning in the presence of external mass transfer resistances, while Kam et al. [183] discussed the analysis of Langmuir-Hinshelwood models to triangular self-poisoning. Reversible Langmuirian inhibition by an impurity was analyzed by Valdman et al. [184] and by Christiansen and Anderson [185], while Lee and Aris [186] discuss a model where the reactant can compete with the poison for the catalytic sites.

The effects of diffusion-influenced poisoning on the apparent kinetics of the main reaction have been investigated by Tartarelli and Morelli [187] and by Hlubucek and Pasek [188]. Approximate methods to solve the diffusion-reaction problem during catalyst poisoning were proposed by Tai and Greenfield [189], and a collocation technique was suggested by Kulkarni and Ramachandran [190]. Dudukovic [191] used a mathematical transformation to reduce the diffusion-reaction problem during catalyst poisoning to an initial value problem. Khang and Levenspiel [192] proposed to replace the intrapellet diffusion-reaction equations which describe catalyst poisoning by a simple n-th order rate form, and discussed the validity of this approach for different poisoning mechanisms.

The case of a catalytically active catalyst poison was investigated by Hegedus [193]: in such cases the original activity level of the catalyst gradually transits to a new activity level which is determined by the catalytic activity of the deposited poison. This problem can be quantified by

$$\partial^2 \psi_A / \partial \eta^2 - h_1^2 \theta \psi_A - h_2^2 (1 - \theta) \psi_A = 0 \tag{80}$$

$$\partial^2 \psi_P / \partial \eta^2 - h_3^2 \theta \psi_P = 0 \tag{81}$$

$$\partial \theta / \partial \tau = -\theta \psi_P \tag{82}$$

$$\psi_A(0, \eta) = f(\eta) \tag{83}$$

$$\psi_P(0, \eta) = 0 \tag{84}$$

$$\theta(0, \eta) = 1 \tag{85}$$

$$\frac{\partial \psi_A}{\partial \eta}(\tau, 0) = \beta_A (\psi_A(\tau, 0) - 1) \tag{86}$$

$$\frac{\partial \psi_P}{\partial \eta}(\tau, 0) = \beta_P (\psi_P(\tau, 0) - 1) \tag{87}$$

$$\frac{\partial \psi_A}{\partial \eta}(\tau, 1) = 0 \tag{88}$$

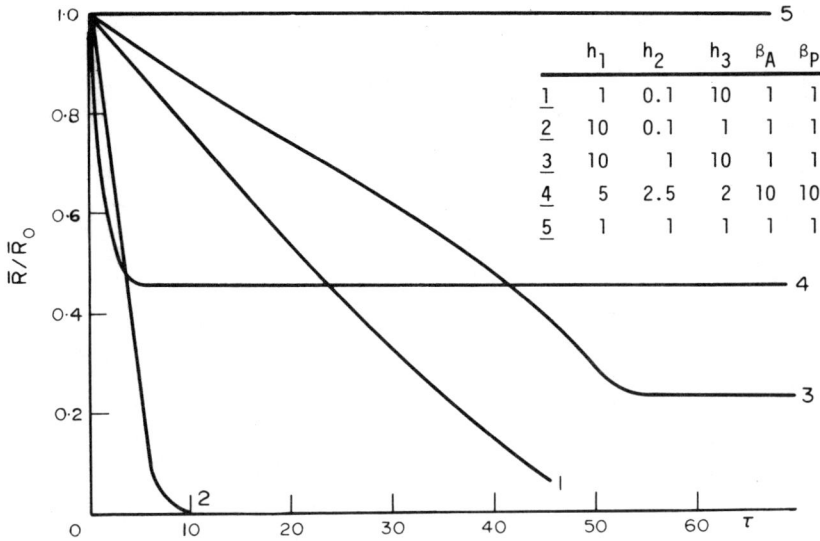

FIG. 36. Poisoning by a poison which is catalytically active, for various Thiele parameters (h) and Biot numbers (β) [193].

$$\frac{\partial \psi_P}{\partial \eta}(\tau, 1) = 0 \qquad (89)$$

where h_1, h_2, and h_3 are the respective Thiele parameters, β_A and β_P the Biot numbers, and θ is the fraction of original surface remaining. The term $h_2^2(1 - \theta)\psi_A$ accounts for the catalytic activity of the poison. For various values of the parameters, Fig. 36 shows the nature of this activity transition problem.

Peculiar phenomena can arise during the poisoning of catalysts with an autocatalytic main reaction: Hegedus et al. [194] found that the isothermal, diffusion-reaction-induced multiplicities which arise in such systems (such as the oxidation of CO over Pt-alumina) can be significantly amplified by pore-mouth poisoning, due to the extra diffusion resistance provided by the poisoned shell. Figures 37 and 38 show these effects; the experimental results agree well with the predictions of a diffusion-reaction model.

To close this part, let us mention the brief review of Schlosser [195] which includes a concise discussion of the single-pellet poisoning problem from the reaction engineering standpoint.

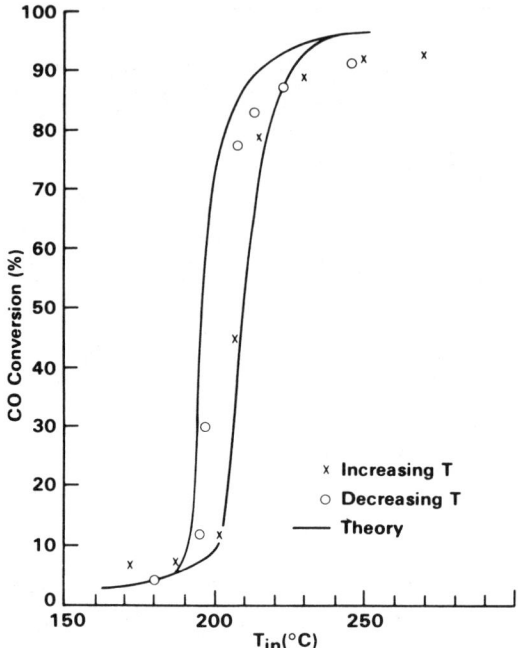

FIG. 37. Isothermal multiple steady states during CO oxidation over fresh Pt-alumina [194].

D. Effects of the Size and Shape of the Catalyst Pellets

The Thiele parameter of the diffusion-influenced poisoning process depends on the size of the catalyst particles, and thus the time scale of the process is significantly influenced by pellet size. In general, smaller pellets tend to collect the poisons faster than larger pellets. The problem is not this simple, however, if the activity history of the catalyst is considered, since the Thiele parameter of the main reaction, too, is influenced by particle size. Thus the pellet size-deactivation rate problem may involve optimum pellet sizes for a given catalytic activity after a certain time elapsed, and this optimum may depend on the mechanism and kinetics of both the main and poisoning reactions.

Anderson et al. [196], for example, found that the sulfur poisoning of Fischer-Tropsch catalysts can be significantly delayed by decreasing the size of the particles. On the other hand, Moseley et al.

EFFECTS OF THE SIZE AND SHAPE

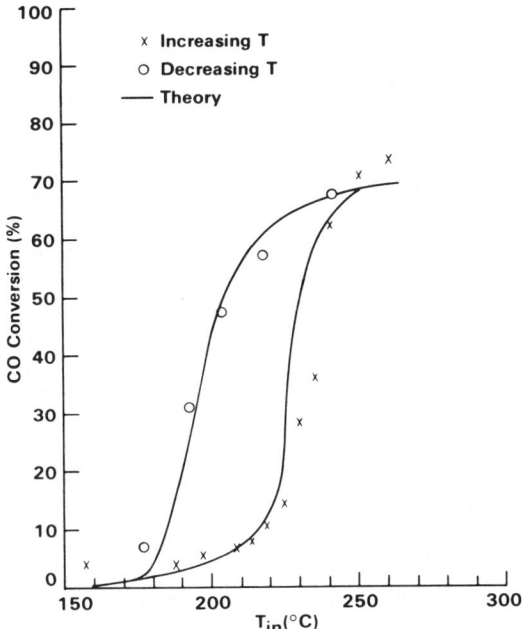

FIG. 38. Isothermal multiple steady states during CO oxidation over poisoned Pt-alumina [194].

[197], who studied the poisoning of a steam hydrocarbon gasification catalyst, found that the poisoning rate fell sharply as the particle size was <u>increased</u> but became independent of particle size for large-enough particles. To complete the picture, El Menshawy et al. [198] discovered an optimum catalyst particle size to maximize the activity of the catalyst during the poisoning of CO oxidation over Al_2O_3-supported NiO by iron carbonyl.

The analysis of the effects of intraparticle diffusion resistances on catalyst poisoning by Masamune and Smith [173] puts pellet size effects into quantitative perspective. For <u>impurity poisoning</u>, they found, the least deactivation will occur if there is a minimum diffusion resistance for the main reactant and a maximum diffusion resistance for the impurity. For a given system this implies the existence of an optimum pellet size for highest activity after a fixed elapsed time. For <u>parallel self-poisoning</u>, again, an intermediate degree of diffusion resistance was found to minimize catalyst poisoning, implying again an optimum pellet size. For <u>series self-poisoning</u>, on the other hand, a catalyst with minimal intrapellet diffusion resistances gives the maximum activity for any process time, suggesting the use of small catalyst pellets.

The shape of the catalyst particles also can influence the time scale of poisoning, due in part to the resulting variations in surface-to-volume ratio of the pellets. For a fixed surface-to-volume ratio, convergent or divergent geometries will result in different poisoning, time scales when compared with slab geometry. For pore-mouth poisoning, Sada and Wen [199] analyzed the effects of pellet geometry on the activity and selectivity of first-order reaction schemes and concluded that the length of time necessary for completely poisoning the pellets increases in the sequence of sphere, cylinder, and slab geometries. Spheres, infinite cylinders, and infinite flat slabs were compared by Hegedus [181] for the case of impurity poisoning with various degrees of transport resistances and, for gas-solid reactions, Gokarn and Doraiswamy [200] concluded that the generalized diffusion length approximation of Aris [201] effectively cancels the shape differences between infinite cylinders, regular cylinders, and flat plates. Thus the actual time scale of the poisoning problem seems to be well approximated by a model of the process in which the Thiele parameters involve the characteristic dimension

$$L = \frac{\text{particle volume}}{\text{external surface area}} \tag{90}$$

for the particular pellet shapes.

E. Pore Structure Effects

The pore structure of the catalyst pellet directly influences the degree of diffusion resistances of both the main and poisoning reactions, and thus, in general, the findings of Masamune and Smith [173] also apply here. However, two special cases merit further discussion.

The first case arises due to the fact that the pore structure of the support determines not only the diffusivity of the gas-phase species but also the internal pore surface area. If the poison precursor can react with the support, different pore structures will have different net effects, even if their diffusive characteristics are invariant. This observation has been exploited in the design of poison-resistant automobile exhaust catalysts as will be discussed in the last section of this paper, and by Rajagopalan and Luss [202] who predicted pore sizes yielding either optimal initial activity or optimal lifetime activity for the catalytic demetallation of hydrocarbons. The protection of the active components of a catalyst from poisoning by subsurface impregnation also relies on this phenomenon, and this area will be discussed later on.

The second important effect of pore structure involves the plugging of pores during catalyst poisoning. This phenomenon often contributes to the catastrophic loss of catalytic activity, since the pores tend to be blocked near their entrance, reinforcing the already severe effects of pore-mouth poisoning. Examples can be taken from catalyst coking (e.g., Richardson [203]), ammonia synthesis (Chen and Anderson [204]), automobile exhaust catalysis (Bomback et al. [205], Chou and Hegedus [206]), or hydrodesulfurization (e.g., Stanulonis et al. [207], and Newson [208]). For complex pore structures, pore plugging may occur even in the absence of diffusional restrictions (Androutsopoulos and Mann [209]). The mathematical modeling of these pore structure changes during catalyst poisoning is similar to the modeling of noncatalytic gas-solid reactions, exemplified by the paper of Ramachandran and Smith [210] and references therein.

For thin plugged crusts, steady-state diffusivity measurements may be insufficiently sensitive; in such cases, transient diffusivity measurements (Chou and Hegedus [206]) proved to be of significant value to quantify the extent of pore plugging. Thus a 6.3-μm thick poisoned and partially plugged shell (Fig. 39) was shown by this technique to have a 26-fold reduction in its diffusive characteristics with respect to the poison-free core of the catalyst pellet.

F. Effects of Catalyst Impregnation Profiles

Intrapellet diffusion effects result in activity and concentration gradients in the catalyst pellets. Therefore, significant effects can be achieved by manipulating the distribution of the active component along the characteristic dimensions of the catalyst pellet.

The literature of catalysts with nonuniform impregnation is large and is in fast growth. Activity gradients have been suggested to manipulate the catalyst's conversion performance (both activity and selectivity) in the absence of poisoning and the resistance of the catalyst to poisoning. Hegedus et al. [211] provided a review of nonuniformly impregnated catalysts, including their methods of preparation. We will only discuss those aspects which pertain to catalyst poisoning.

The earliest examples appear in the patent literature. For example, Michalko [212] and Hoekstra [213] patented noble metal-alumina catalysts with a subsurface impregnation of the active components. In the case of pore-mouth poisoning, the poisons react with the bare, unimpregnated support surrounding the active portion of the pellet and thus would delay its poisoning.

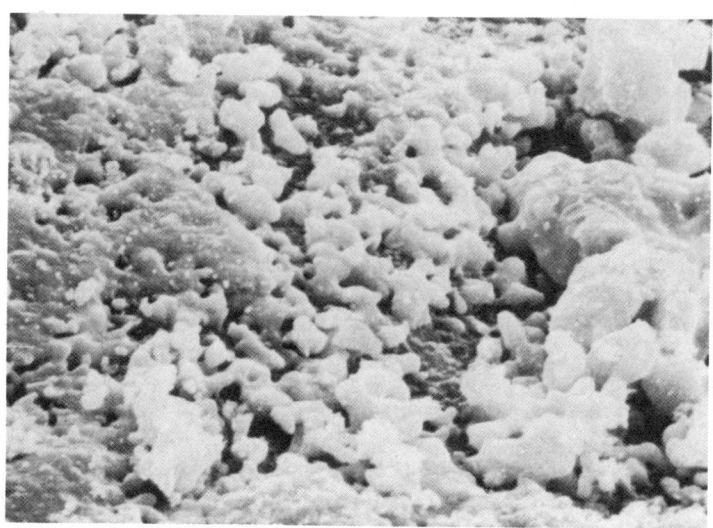

FIG. 39. Scanning electron microscopic photograph of the outer surface of a partially plugged catalyst pellet [206].

The first quantitative analysis of the effects of impregnation profiles on catalyst poisoning is due to Shadman and Petersen [214]. They investigated the <u>series self-poisoning scheme</u>

$$A \to B \to C \qquad (91)$$

where B is the desirable product and C is a strongly chemisorbed, poisonous by-product. Catalyst impregnation profiles were quantified by

$$a = a_0 \xi^\alpha \qquad (92)$$

where ξ is the dimensionless distance from the pellet's center, a is the local active surface area (in, e.g., cm^2/cm^3 pellet), and α is a parameter the variation of which results in various impregnation profiles (Fig. 40). For a fixed degree of intrapellet diffusion resistances (in their example, $h_A = 5$), initial activity can be traded for increased poison resistance by introducing activity profiles which gradually decrease toward the pellet's center (Fig. 41). The decline of initial activity with increasing α is due to the fact that a_0 was kept constant and

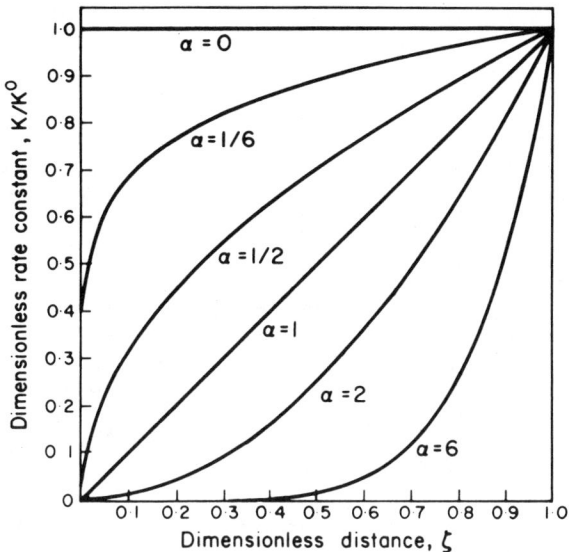

FIG. 40. Catalyst impregnation profiles [214].

thus the total amount of active component declined with increasing α. Constant total catalyst loading results in a simultaneous increase in both initial activity and durability for this case of series self-poisoning, as Corbett and Luss [215] demonstrated.

DeLancey [216] employed Pontryagin's minimum principle to calculate the optimum distribution of the active component in a catalyst pellet which is being deactivated by uniform poisoning. Full activation to a fractional depth, with an inert core, was found to be optimal.

Wei and Becker [217] considered the oxidation of CO over Pt-alumina. For this autocatalytic reaction which can be of negative order, a thick active layer was proposed, as opposed to thin layers proposed by Minhas and Carberry [218] for positive-order reactions. Deeper impregnation depths would result not only in increased activity for negative-order reactions but also in increased resistance to impurity poisoning.

Corbett and Luss [215] provided a detailed analysis of impurity poisoning and series self-poisoning for catalysts with nonuniform impregnation gradients. For positive-order reactions with impurity poisoning of the pore-mouth type, they concluded that catalysts which

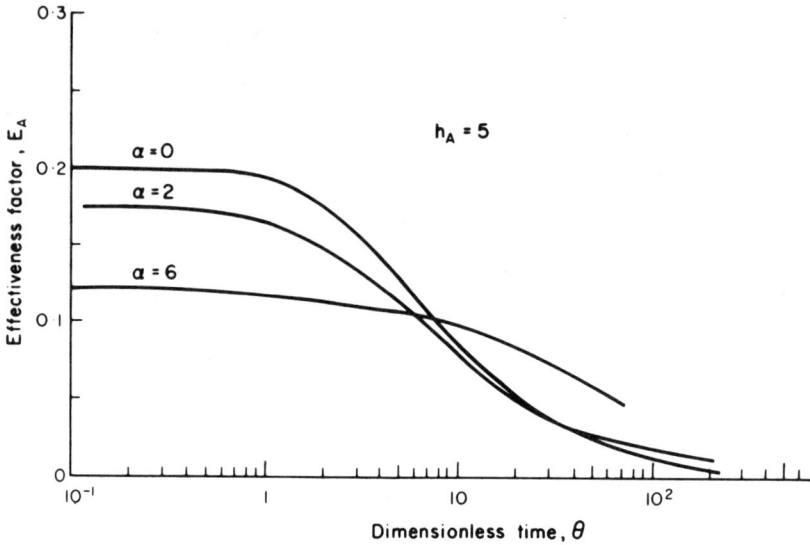

FIG. 41. Effects of catalyst impregnation profiles on the poison resistance of a catalyst during series self-poisoning [214].

concentrate the active component near the interior of the pellet are more poison-resistant, while catalysts with the active component near the pellet's exterior have higher initial activity but deactivate faster. Thus the impregnation strategy depends on the intended process lifetime of the catalyst. There is a point in time where the two strategies are equivalent; for shorter process times, catalysts with exterior enrichment are preferable, while for longer process times the choice is interior enrichment. We will see later on how important these considerations can become in the design of poison-resistant catalysts for automobile emission control, where the preservation of both initial activity and activity after a specified process time are of paramount interest, coupled with the need of conserving the expensive noble metal components (Hegedus and Summers [220], Summers and Hegedus [221], and Hegedus et al. [222].

Becker and Wei [219] compared the performance of four catalyst pellet designs during <u>pore-mouth poisoning</u> with first-order reaction. (The problem is applicable to automobile exhaust catalysis, since even the autocatalytic oxidation of CO over Pt reverts to positive order in the presence of the significant transport resistances which prevail in

CATALYST IMPREGNATION PROFILES 71

warmed-up catalytic converters.) They placed an impregnated ring beginning either at the catalyst's outer surface ("egg shell"), below the outer surface ("middle egg white"), to the center ("egg yolk"), and uniformly distributed over the entire pellet. For the particular band widths considered, they concluded that maximum activity is accomplished by different designs depending on the duration of the poisoning experiment (i.e., the penetration depth of the pore-mouth-poisoned shell): for fresh performance the "egg shell" design is preferable, while the "middle egg white" configuration was proposed for long-term durability, suggesting the subsurface impregnation of a catalytically active band of finite thickness (Fig. 42), in agreement with the patented configuration of Hoekstra [213]. These calculations considered only the case where a fixed fraction (one-third) of the volume of the support sphere is impregnated, resulting, e.g., in a fixed impregnation depth for the "egg shell" profile. We will discuss later on how "egg shell" profiles can be designed for an appropriate thickness taking into account the extent of poisoning after a fixed elapsed time of poison exposure.

Various configurations can be envisioned along the above lines. Thus Wolf [223] and Polomski and Wolf [224] explored subsurface-impregnated designs where the outer inert layer is composed of a different pore structure than the inner core. Different metals (Pt, Pd, Rh) can be advantageously impregnated in different layers of appropriate sequence and thickness (Summers and Hegedus [221], Hegedus et al. [222]); these ideas will be discussed later on.

Beyond the macroscopic distribution of the active components in the catalyst pellets, localized effects also may be of significant importance. Thus Wolf [225] proposed to compose a catalyst pellet of a homogeneous mixture of impregnated and inert support grains, the two grain types having different effective diffusivities. For pore-mouth poisoning, such diluted catalysts were computed to show both higher activity after a fixed time on stream and a slower rate of deactivation.

The calculations of Becker and Wei [219] employed a steady-state analysis, computing catalyst performance as a function of catalyst temperature. In automobile exhaust service, the dynamics of the system becomes very important, since the problem is posed in the emissions-time domain.

Oh et al. [226] recomputed the Becker-Wei problem in a dynamic mode, and analyzed the performance-time characteristics of various Pt impregnation profiles before and after pore-mouth poisoning. Surface-impregnated Pt bands of appropriately designed thickness appeared to provide a good compromise between initial and final activity. The in-

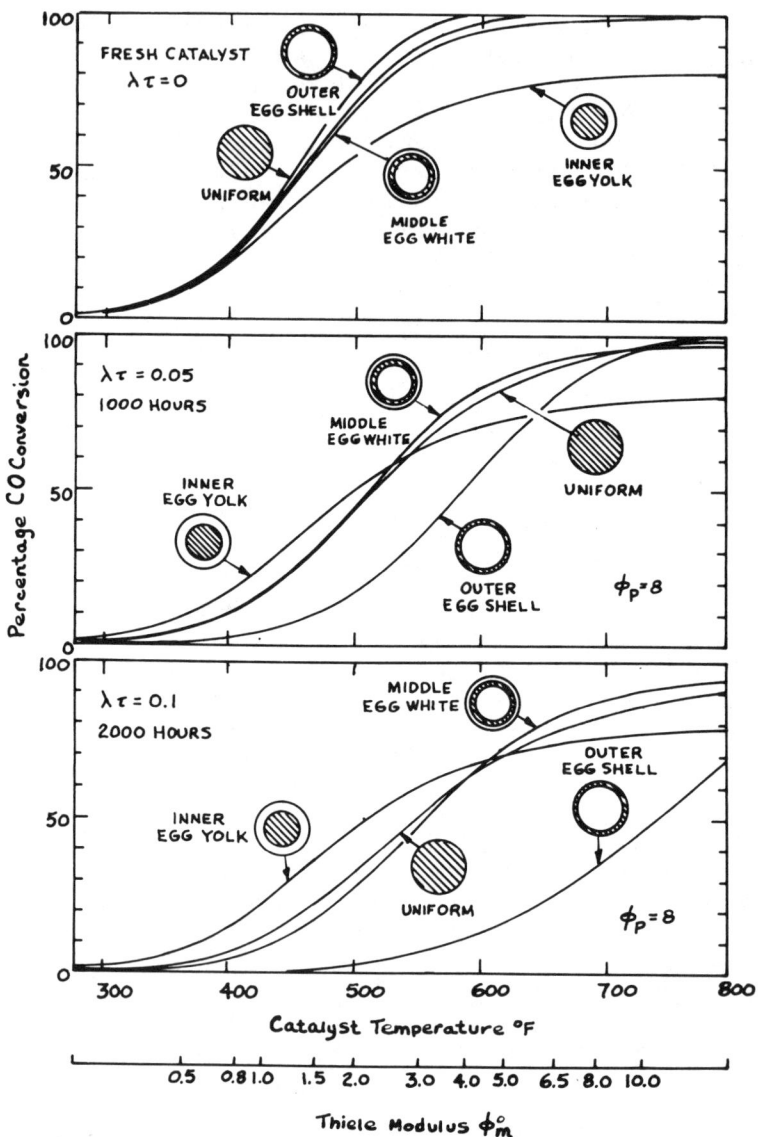

FIG. 42. Effects of various catalyst impregnation strategies on the poisoning of the CO oxidation reaction by an impurity [219].

SELECTIVITY PROBLEMS

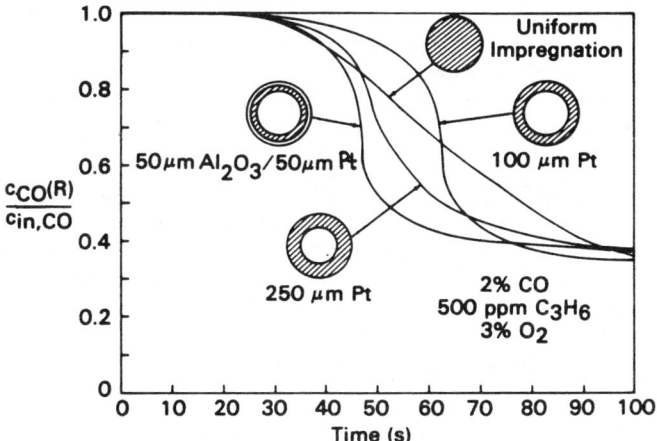

FIG. 43. Lightoff behavior after $50\,\mu$m poison penetration for various impregnation strategies [226].

teresting finding of this analysis was that by the time catalyst lightoff occurred, the intrapellet temperature profiles were reasonably flat so that the Becker-Wei analysis is appropriate for ranking the various impregnation strategies. Figure 43 shows typical lightoff times for a few pellet designs after the outer $50\,\mu$m of the pellets was poisoned by phosphorus.

G. Selectivity Problems

There are two kinds of selectivity problems in catalyst poisoning which need to be differentiated. In one case, competing reactions run on the same kind of sites, and the poisoning of these sites will change the selectivity. In the second case, different kinds of sites promote different reactions, and catalyst poisoning will change the selectivity due to the fact that these different sites are poisoned at different rates. These aspects, as far as the chemistry is considered, were discussed in the previous section, so we will restrict our attention here to the effects of diffusion resistances on the poisoning of such selectivity systems.

In a broader sense, catalyst poisoning itself represents a selectivity problem, where the design of poison-resistant catalysts involves the selective suppression of the rate of the poisoning events. In a narrower sense, as it is applied here, we will focus our attention on the competition of two or more main reactions as affected by poisoning.

An excellent discussion of diffusion-influenced selectivity problems was provided by Wheeler [156]; poisoning effects on selectivity can be best discussed by first referring the reader to the above paper.

Sada and Wen [199] investigated the effects of pore-mouth poisoning by an impurity on the selectivity of independent, parallel, and consecutive reactions for three different catalyst geometries. The following mechanisms were considered.

Independent reactions, B desired:

$$A \xrightarrow{k_1} B \tag{93}$$

$$E \xrightarrow{k_2} F \tag{94}$$

Parallel reactions, B desired:

$$A \begin{array}{c} \nearrow^{k_1} B \\ \searrow_{k_2} F \end{array} \tag{95}$$

Consecutive reactions, B desired:

$$A \xrightarrow{k_1} B \xrightarrow{k_2} F \tag{96}$$

Two cases were considered by Sada and Wen [199]: a case where only the desired reaction is being poisoned and a case where both the desirable and undesirable reactions are poisoned.

For the first case the reduction of selectivity as a function of the fraction of the pellet poisoned is shown in Figs. 44, 45, and 46 for independent, parallel, and consecutive reactions, respectively. In the case of independent reactions, the numerical value of the parameter

$$\beta = \sqrt{k_2 D_A / k_1 D_E} \tag{97}$$

SELECTIVITY PROBLEMS

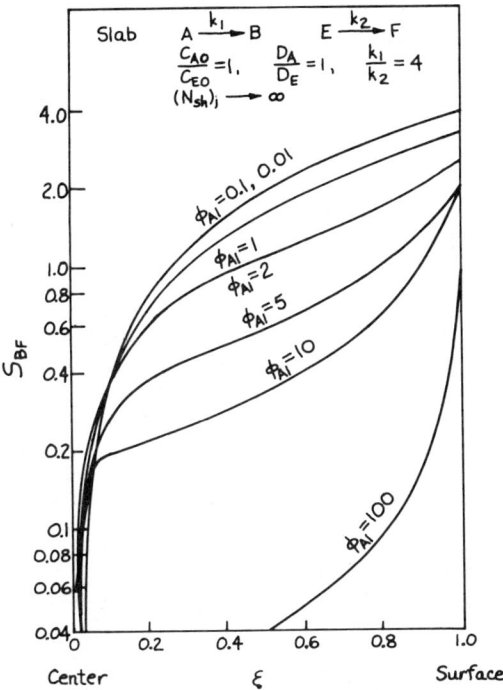

FIG. 44. Reduction in selectivity as a function of active region remaining in a catalyst slab for independent reactions [199].

determines the optimum strategy; for $\beta > 1$ an optimum degree of internal resistances exists, quantified by the Thiele parameter

$$\Phi_{A1} = R\sqrt{k_1/D_A} \tag{98}$$

while for $\beta < 1$ the smallest possible value of Φ_{A1} will yield the best selectivity. For parallel and consecutive reactions, the smallest value of Φ_{A1} is to be chosen for best selectivity.

For the second case where both the desired and undesired reactions are poisoned, pore-mouth poisoning will either increase the

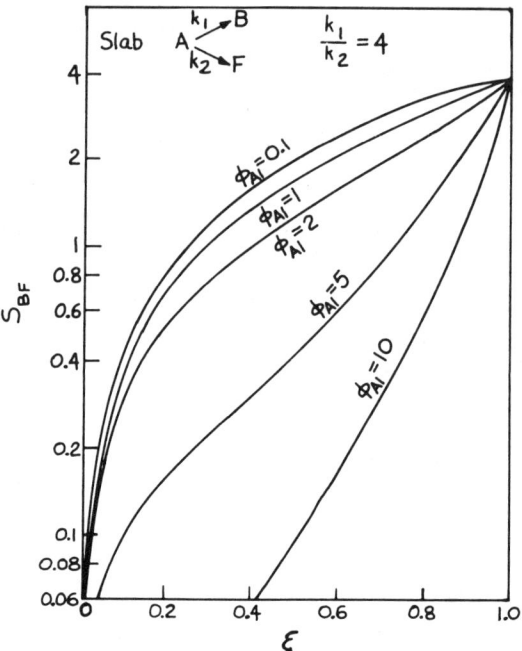

FIG. 45. Reduction in selectivity as a function of the fraction of activity remaining in a catalyst slab for parallel reactions [199].

selectivity or leave it constant, and the largest possible value of Φ_{Al} is indicated to achieve best selectivity.

The selectivity change during the hydrogenation of acetylene over supported Ni was analyzed by Komiyama and Inoue [227], both experimentally and theoretically. Uniform poisoning, pore-mouth poisoning, and core poisoning were considered with Langmuir-Hinshelwood rate expressions. The analysis included the case of pore-mouth poisoning where the poisoned shell retained a residual activity. The reaction scheme is

$$C_2H_2 \xrightarrow{k_1} C_2H_4 \xrightarrow{k_2} C_2H_6 \tag{99}$$

with C_2H_4 as the desired product. The calculations for a shell-poisoning model agreed well with the observed selectivity changes.

SELECTIVITY PROBLEMS

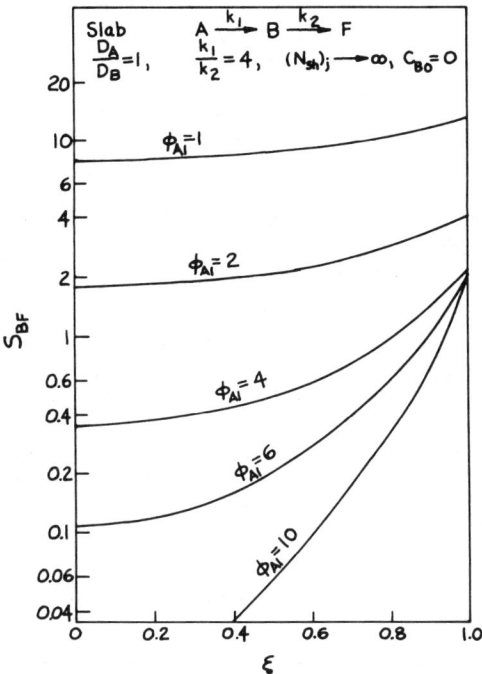

FIG. 46. Reduction in selectivity as a function of the fraction of activity remaining in a catalyst slab for consecutive reactions [199].

Corbett and Luss [215] analyzed the effects of impurity poisoning and series self-poisoning on the selectivity of the system

$$A \xrightarrow{k_1} B \xrightarrow{k_2} C \qquad (100)$$

with B as the desirable product, with emphasis on the effects of different catalyst impregnation profiles, pointing out the possible necessity of compromises between selectivity and poison resistance when an appropriate impregnation profile is to be selected.

Karanth and Luss [228] analyzed the effects of uniform poisoning on the selectivity of system (96) toward B in a diffusion-influenced catalyst pellet, with Langmuir-Hinshelwood kinetics of both Reactions (1) and (2). Poisoning of this nature was shown to enhance the selectivity by slowing down the reactions and so reducing the deleterious effects of intraparticle diffusion limitations, thus explaining the effects of selectivity moderators often employed in diffusion-influenced industrial reactor operations.

Other interesting features of the selectivity system shown in Eq. (96) were pointed out by Pareja and Luss [229] for the case where different numbers of catalytic sites are occupied by the various reacting species. In the absence of diffusion resistances, the selectivity is a monotonic function of θ, the fraction of sites poisoned. However, when diffusional resistances exist, the selectivity may attain extremal values as a function of θ. Figure 47 illustrates this phenomenon, where ψ_A is the Thiele parameter of the poison-free pellet and $a = n - m$ (n is the number of sites occupied by B, and m is the number of sites occupied by A).

The second kind of selectivity problem involves the <u>poisoning of multifunctional catalysts</u>. Lee and Butt [230] analyzed the system

$$A \xrightarrow{S_1} B \xrightarrow{S_2} C \qquad (101)$$

where the two reactions occur on separate sites. Impurity poisoning and parallel and series self-poisoning were considered, both in the presence and absence of diffusion limitations. In the impurity poisoning case, poison L deactivated site S_1 and poison M deactivated site S_2; for parallel self-poisoning, A deactivated S_1 and B deactivated S_2, while for series self-poisoning, S_1 was deactivated by B and S_2 by C. The analysis shows how diffusion effects can benefit both catalyst activity and life, and how such complex selectivity events can be understood and manipulated, with significant implications to industrial catalytic practice.

H. Nonisothermal Catalyst Pellets

Since poison gradients modify the activity gradients in catalyst pellets, they are also expected to have an effect on intrapellet temperature gradients. These modified intrapellet gradients, in turn, may modify not only the rates of the main reactions but also those of the poisoning reactions. While catalysts with uniform activity profiles rarely develop significant intrapellet temperature gradients since the main resistance to heat transfer often resides in the gas phase (Carberry [231]), significant intrapellet gradients may develop as a result of catalyst poisoning which often have to be taken into consideration in predicting the activity history of the pellet during the course of deactivation.

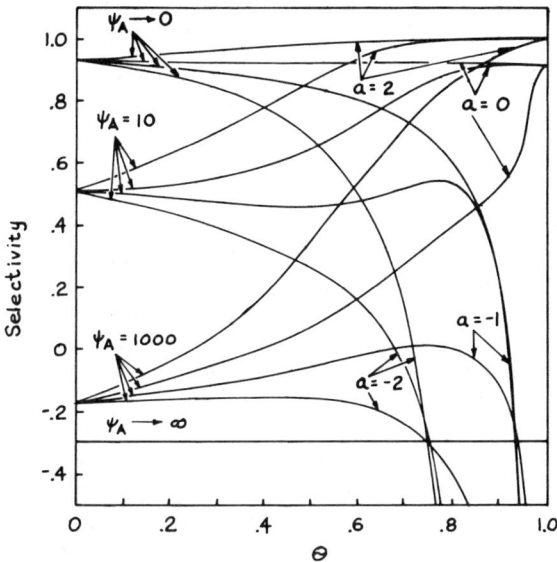

FIG. 47. Effects of impurity poisoning on the selectivity of a consecutive reaction A → B → C. ψ_A is the Thiele parameter, A requires m = 4 sites, B requires n sites, and a = n - m [229].

The complexity of the problem is well illustrated by the work of Sagara et al. [232] who investigated the effects of an exothermic main reaction on parallel and series self-poisoning. Unusual activity profiles were computed for parallel self-poisoning at finite poisoning times, due to the opposing effects of temperature and concentration gradients, resulting in maximum activity below the pellet's outer surface. On the other hand, for series self-poisoning the concentration and temperature gradients are complimentary, resulting in monotonically declining activity toward the pellet's center.

Ray [233] analyzed the effects of pore-mouth poisoning on intrapellet temperature gradients during an exothermic reaction, and detected thermally induced multiple steady states at intermediate degrees of poisoning.

Koh and Hughes [234] conducted experimental temperature measurements within a catalyst pellet during impurity poisoning (by traces of O_2) of ethylene hydrogenation over Pt-alumina. They concluded that while the uniform pellet temperature assumption is justified for the active pellet, for the poisoned case the intraparticle temperature rise may reach 50% or more of the extraparticle temperature rise and therefore cannot be neglected. Butt et al. [235] have subsequently shown that the relative importance of intraparticle temperature rise with respect to the extraparticle temperature rise increases with progressive degrees of thiophene poisoning during the exothermic catalytic hydrogenation of benzene (Fig. 48), and that the expression derived by Carberry [236] is capable of quantitatively predicting these effects.

In light of the changing relative importance of extra- and intraparticle temperature gradients with the extent of poisoning, it would be interesting to see how the findings of Sagara et al. [232] and Ray [233] would modify if extrapellet temperature gradients were allowed (both assumed no extrapellet temperature gradients). A refinement of Carberry's estimation technique was recently published by Oh et al. [237] and by Lee [238].

The transient response of poisoned catalyst pellets was analyzed by Lee et al. [239] and Downing et al. [240]; these and other papers on the effects of poisoning on catalyst and reactor dynamics were reviewed recently by Butt [10] and thus will not be discussed here. However, it is worth noting that, just as their steady-state behavior, the dynamic characteristics of catalysts can also be significantly influenced by poisoning.

4
Design of Poison-Resistant Catalysts: A Case History

The careful analysis of complex catalyst poisoning events and the resulting understanding of the mechanism of poisoning can significantly improve the design of commercial catalysts. While it is recognized that probably the bulk of catalyst development work is empirical in nature, we nevertheless would like to show an example where a thorough analysis paid off. This example is taken from automobile exhaust catalysis which, in terms of catalyst sales, is apparently the largest-scale catalytic process (Burke [241]).

The exhaust of gasoline-powered automobiles contains numerous catalyst poisons, such as various volatile forms of Pb, P, and S compounds. The nature of these poisons and our ability to control their concentration or their effect on the catalyst had far-reaching consequences to the chemical industry (declining sales of tetraethyl-Pb additives due to their removal from the fuel), the oil industry (refining gasoline to higher octane numbers due to the removal of Pb-containing octane boosters), and to the automobile industry (redesigning certain engine components for operation on lead-free fuel). In addition, as we will discuss, the chemical composition and physical design of the catalysts are dominated by poisoning considerations.

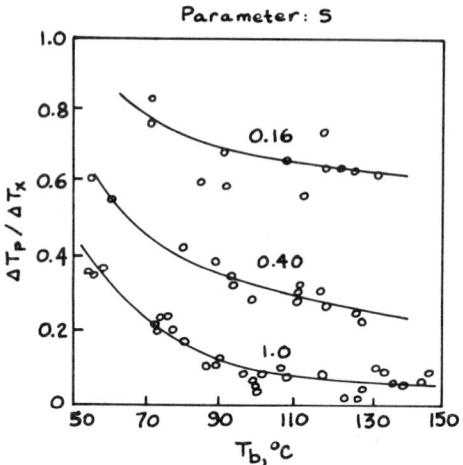

FIG. 48. Effects of deactivation on the relative magnitudes of inter- and intrapellet temperature gradients at various degrees of poisoning (S is the fraction of initial activity [235]).

The so-called lead-free fuel contains only small traces of Pb (about 0.00045 g/L) and P (about 0.00005 g/L), but significant quantities of S (about 0.080 g/L). An additional and most important source of poisons is the P-content of the lubricating oil (about 0.1 g/L): P-free oils of sufficient performance have not yet been developed.

Sulfur compounds burn to SO_2 in the engine, resulting in approximately 20 ppm (by volume) of SO_2 in the exhaust. Base metal catalysts of sufficient intrinsic activity are severely poisoned by sulfur compounds, and the removal of sulfur traces from gasoline appears to represent an as of yet unsurmounted obstacle. Thus, in order to operate in sulfur-containing exhaust, noble metals had to be selected for automobile exhaust cleanup. A recent discussion of these developments was provided by Hegedus and Gumbleton [242], while a review of the poisoning of automobile exhaust catalysts by Shelef et al. [50] may be consulted for many of the chemical details involved.

The first generation of automobile exhaust catalysts was developed for the oxidation of CO and hydrocarbons to CO_2 and H_2O. Large-scale development work established Pt and Pd as the active components, supported over stabilized γ-alumina. The support consists either of porous alumina pellets of about 0.32 cm diameter or of alumina-coated monoliths. We will restrict our discussion to the former.

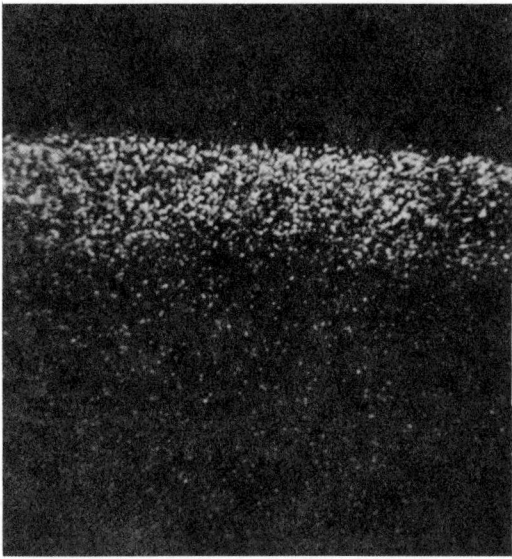

FIG. 49. Cross section of a phosphorus-poisoned catalyst pellet (by electron microprobe [220]).

Two kinds of significant poisoning phenomena were observed over noble metal catalysts: a fast, reversible inhibition by SO_2 (and, to a certain extent, by halogen compounds from traces of Pb scavangers), and a slow, essentially irreversible poisoning by P and Pb compounds.

Electron microprobe studies on sectioned, aged catalyst pellets (Fig. 49) revealed that the poisons penetrate the pellets in the form of a sharply defined shell, progressing toward the center of the pellets as time elapses. The sharpness of these poison bands indicated that they are diffusion limited, and their analysis by electron microprobe showed that the position of the leading edge is determined by P poisoning. Thus P from the lubricating oil proved to be the main poison once Pb was reduced to traces in the fuel.

Quantitative electron microprobe scans determined that the saturation concentration of P in these poisoned bands is close to a monolayer coverage of the alumina surface (Fig. 50), suggesting that the poison precursor (H_3PO_4) readily reacts with the alumina support. This proved to be of crucial importance in designing poison-resistant catalysts, since theoretical calculations (Hegedus [181]) indicated significant advantages of employing the support as a poison-getter to protect the still unpoisoned noble metals beyond the poisoned zone.

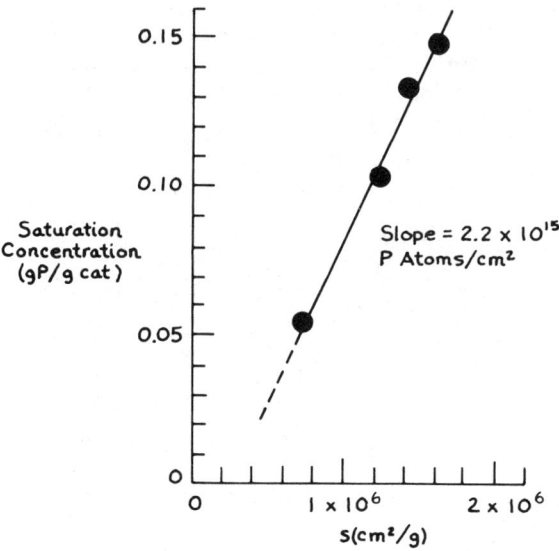

FIG. 50. Phosphorus saturation concentration as a function of support surface area (from electron microprobe experiments [243]).

Kinetic studies revealed that at these high temperatures (about 500-550°C) the main reactions (oxidation of CO and of most of the hydrocarbon species) are almost completely limited by diffusion across the poisoned shell, and can be well approximated by first-order kinetics.

Diffusivity measurements and diffusion-limited reactivity measurements showed that in the normal modes of catalyst operation, the poisoned shell has the same diffusive characteristics as the bulk of the unpoisoned pellet; this can be explained by the monolayer-like poison coverage in the poisoned shell so that the transport pores of the support are not plugged.

To test this diffusion-limited poisoning mechanism a four-tray plug-flow reactor was built (Fig. 51), and catalyst samples were taken from the trays at various times during an accelerated P-poisoning experiment. The pellets were analyzed both for poison content and poison penetration as a function of elapsed poisoning time.

The mathematical model of this experiment involves an isothermal, diffusion limited, plug-flow scheme, with shell-progressive poisoning of the individual pellets:

$$\frac{\partial \psi}{\partial \xi} + \frac{\partial \theta}{\partial \tau} = 0 \quad \text{(gas-phase balance)} \tag{102}$$

CASE HISTORY 85

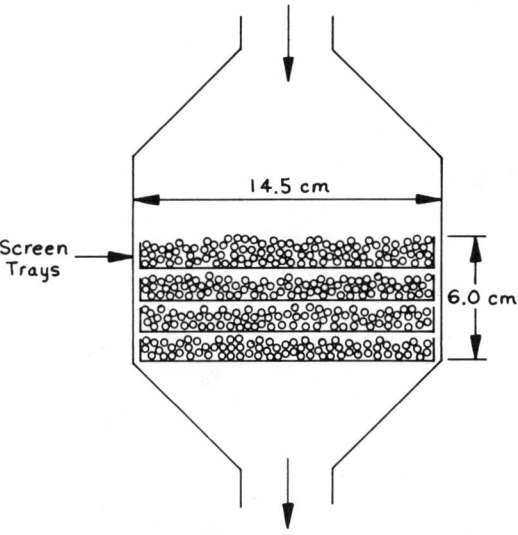

FIG. 51. Four-stage reactor for catalyst poisoning experiments in automobile exhaust [243].

$$\frac{\partial \theta}{\partial \tau} + \Phi(\theta)\psi = 0 \quad \text{(solid-phase balance)} \tag{103}$$

$$\psi(0, \tau) = 1 \tag{104}$$

$$\theta(\xi, 0) = 0 \tag{105}$$

with

$$\Phi(\theta) = \frac{\alpha}{\dfrac{1}{Bi} + \dfrac{1 - (1 - \theta)^{1/3}}{(1 - \theta)^{1/3}}} \tag{106}$$

The physical constants of the system, determined independently, are collected in α, Bi, and τ:

$$\alpha = (1 - \epsilon)\frac{D_{P,\,eff} L}{R^2 v} \tag{107}$$

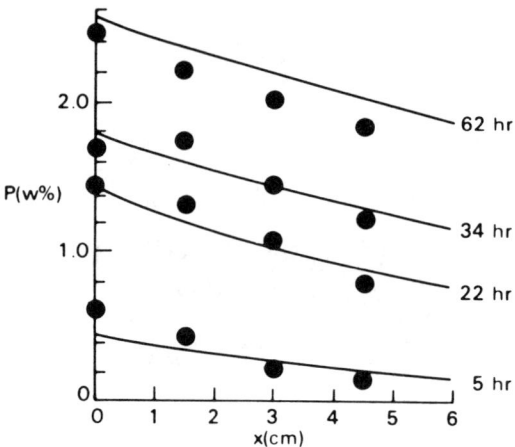

FIG. 52. Comparison of measured phosphorus concentration profiles with the predictions of a pore-mouth poisoning model [243].

$$Bi = \frac{k_m R}{D_{P,eff}} \quad \text{(Biot number)} \tag{108}$$

$$\tau = t\left(\frac{1}{1-\epsilon}\frac{c_o v}{c_s L}\right) \tag{109}$$

Excellent agreement between this model and the experiment was found both for the phosphorus profiles in the bed and for the total phosphorus content of the bed (Figs. 52 and 53), providing proof for the diffusion-controlled mechanism of poison collection. Thus the stage was set for exploring a redesign of the catalyst for improved poison resistance.

Experiments on production-scale catalytic converters revealed that, on the time scale of the poisoning process (80,000 km of driving), the pellets move around sufficiently so that the converter is well mixed and each pellet is poisoned to an equivalent extent. This allowed us to focus our attention on the design of an individual catalyst pellet.

The time-dependent progression of the poisoned shell into the catalyst pellet was described by the shell-progressive model of Carberry and Gorring [174], neglecting the effects of external transport

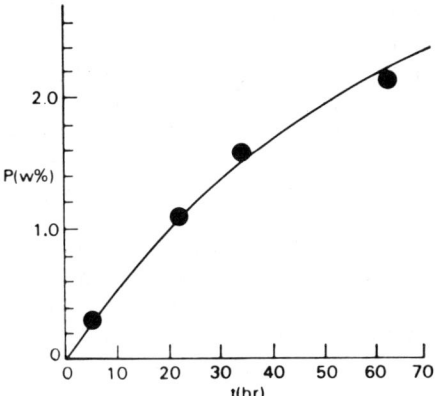

FIG. 53. Comparison of measured phosphorus accumulation rate with the prediction of a pore-mouth poisoning model [243].

resistances and assuming complete diffusion control of the poisoning process across the already poisoned shell, resulting in

$$t = \alpha(-\xi^3 + \frac{3}{2}\xi^2) \qquad (110)$$

where

$$\alpha = \frac{c_{P,s} R^2}{3 D_{P,eff} c_{P,o}} \qquad (111)$$

and

$$\xi = 1 - \frac{r_P}{R} \qquad (112)$$

the fractional poison penetration depth. Thus the poison penetration depth after a given elapsed time t is controlled by two design parameters: the saturation concentration $c_{P,s}$ of the poison (which is related to the surface area of the support) and the effective diffusivity $D_{P,eff}$ of the poison precursor (which is related to the pore structure of the support). The inlet poison concentration $c_{P,o}$ and the radius of the catalyst pellet R are constant: the poison concentration in the exhaust is given, and the pellet radius is determined by pressure drop considerations.

It has also become evident that the impregnation depth of the noble metals is another important design parameter: it has to be designed so that the poisoned front does not pass through it before the intended design life of the catalyst. On the other hand, a too deep noble metal impregnation depth would cause excessive diffusion limitations, resulting in poor catalyst utilization.

The activity of the partially poisoned catalyst is determined by the thickness of the poisoned shell and by the diffusivity of the reactants across it. Thus, just as in the poisoning process, the resulting catalytic activity is also a function of the support's surface area (which controls poison penetration depth) and pore structure (which controls diffusivity).

The fractional activity of a catalyst after a given time of poisoning can be expressed as

$$\epsilon(t) = \frac{1}{1 + Bi_A \frac{\xi}{1-\xi}} \qquad (113)$$

where

$$Bi_A = \frac{k_{A,m} R}{D_{A,eff}} \qquad (114)$$

the Biot number.

Figure 54 shows the parameterized solution of these design equations, using $D_{A,eff}$ as the indication of pore structure and $c_{P,s}$ as an indication of support surface area. The product of $tc_{P,o}$ is fixed, and the straight lines correspond to constant poison penetration depths (i.e., constant noble metal impregnation depths). The curved lines are the contour lines for a constant fractional final activity.

Point A shows one of the first production catalysts. If its performance is to be improved, we have to move in the $D_{A,eff}$ vs $c_{P,s}$ plane toward higher final efficiencies. This can be accomplished by simultaneously increasing both diffusivity and surface area and impregnating the noble metals to the depth indicated for the resulting point.

The validity of the above scenario has been proven in a series of simple experiments. For example, the importance of pore structure was demonstrated by preparing a series of alumina pellets of

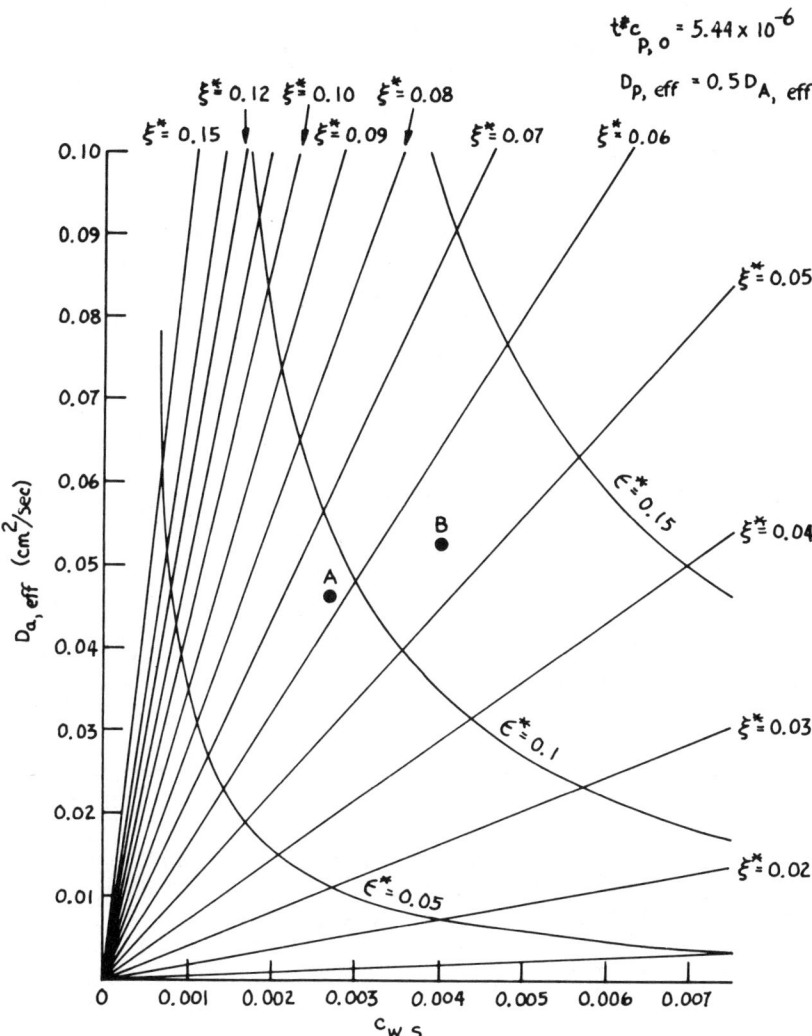

FIG. 54. Effects of catalyst design parameters (diffusivity, surface area, impregnation depth) on the performance of a catalyst pellet during pore-mouth poisoning [220].

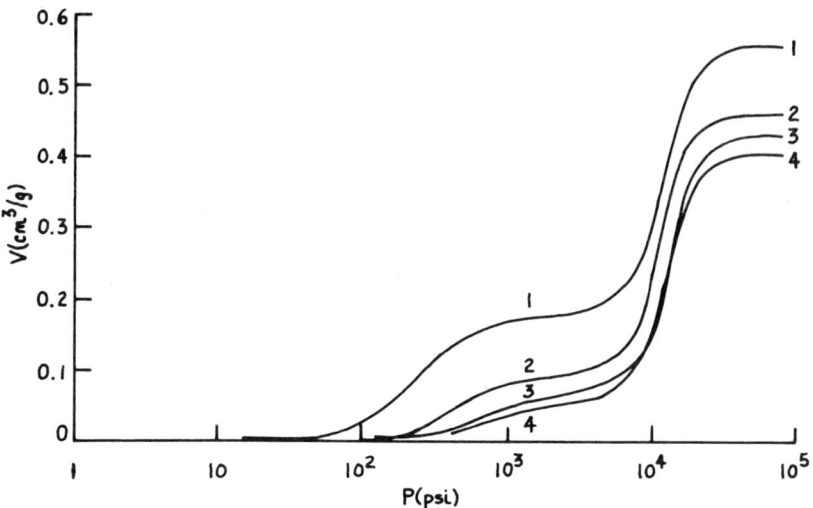

FIG. 55. Hg porosimetric curves of a series of catalysts with varying pore structure [220].

gradually decreasing density (i.e., gradually increasing diffusivity) while keeping the surface area unperturbed. Porosimetric curves of these supports are shown in Fig. 55. The supports were uniformly impregnated and poisoned in an accelerated experiment. As Fig. 56 shows, the low-density structures show higher activity after poisoning, reinforcing the importance of the diffusive transparency of the poisoned layer.

Using these ideas, a new class of low-density aluminas evolved containing larger macropore volumes and radii, larger surface areas, and correctly designed noble metal impregnation depths when compared to early oxidation catalysts (e.g., Hegedus and Summers [244]). Accelerated catalyst poisoning experiments (Fig. 57) and vehicle fleet tests (Adomaitis et al. [245]) proved their viability, and such low-density catalysts are now in large-scale commercial service.

Further analysis of the problem opened the way for additional improvements. These arose from the observation that Pt and Pd, when employed alone, have quite different poisoning characteristics: while Pd completely loses its activity once the poisoned shell penetrates

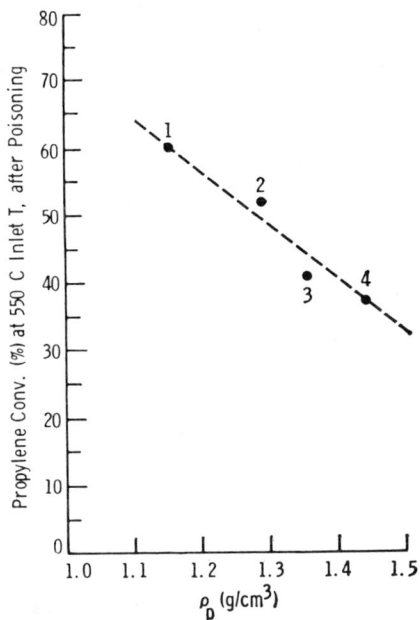

FIG. 56. Propylene conversions after poisoning for catalysts of differing pore structures [220].

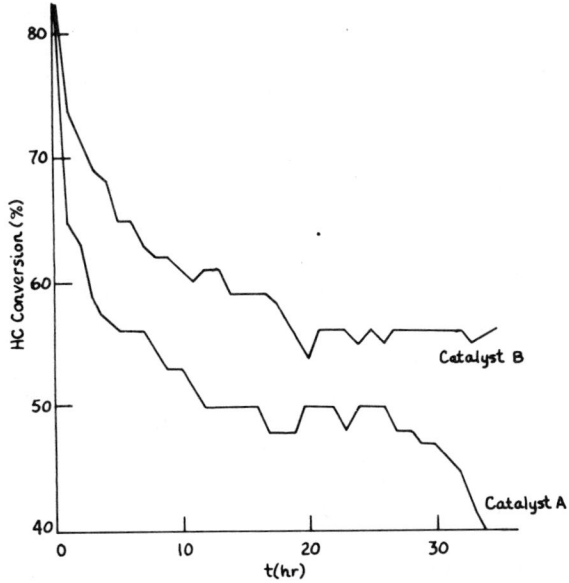

FIG. 57. Comparison of the poison resistance of two catalysts (see Fig. 54) [220].

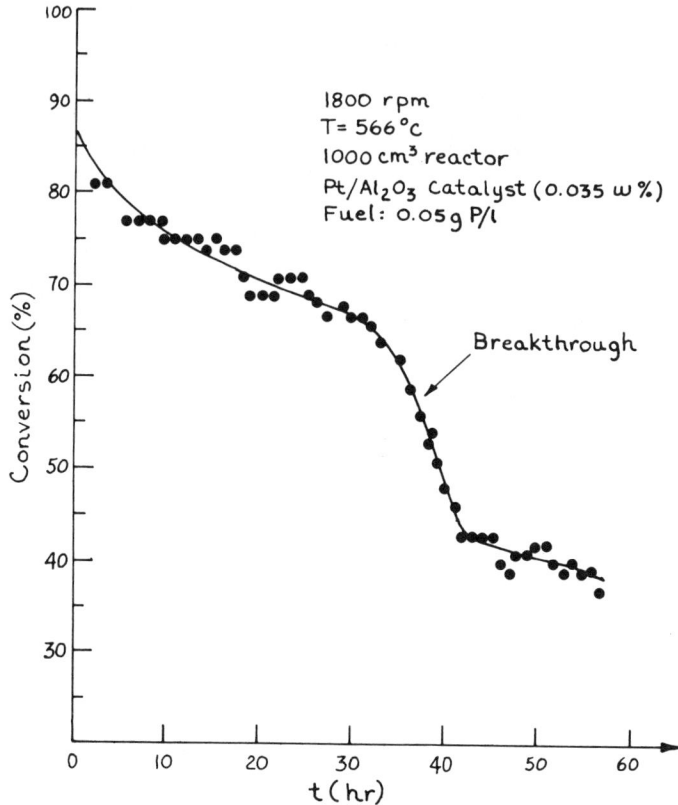

FIG. 58. Poisoning experiment showing the effects of the breakthrough of a poison front across the impregnated shell in the pellets [220].

through the Pd-impregnated shell, Pt retains a certain fraction of it; i.e., it cannot be completely poisoned in automobile exhaust. This was demonstrated in accelerated poisoning experiments on Pt-alumina catalysts: as the poisoned front passes beyond the Pt-impregnated zone (Fig. 58), the activity does not drop to zero.

These differences between Pt and Pd were exploited by impregnating the metals into separate bands, the more poison-resistant Pt closer to the outer pellet surface and the more poison-sensitive Pd in a band below it. The results (Fig. 59) indicate a significant further improvement in poison resistance, partly due to the fact that the Pd is protected from poisoning, and partly that Pt and Pd, when co-

CASE HISTORY

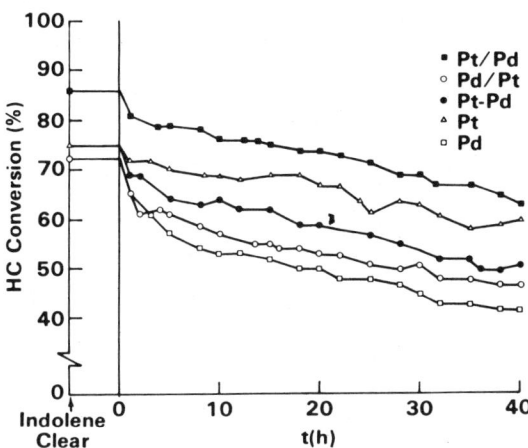

FIG. 59. Effects of Pt and Pd impregnation strategies on the poison resistance of automobile exhaust oxidation catalysts [221].

impregnated within the same band, form an alloy with Pd on its outer surface (this was elegantly proven later by the experiments of Chen and Schmidt [246]), so that coimpregnated Pt-Pd catalysts tend to poison with the characteristics of Pd, the partially poison-resistant character of Pt thus being suppressed by alloying with Pd.

One alternate possibility, namely the impregnation of both Pt and Pd below the outer surface of the catalyst pellet, was rejected on two grounds: low initial activity (suppressed by the diffusion lid imposed by the inert shell) and the increased sensitivity of Pt-Pd alloys to thermal sintering.

The separation of Pt from Pd in separately impregnated rings (Summers and Hegedus [247]) thus provided us with yet another useful catalyst design parameter, illustrating how the essential features of a complex poisoning process, once identified, can be utilized toward the design of improved catalysts.

Similar strategies were employed to improve the poison resistance of three-way catalysts which simultaneously oxidize hydrocarbons and CO, and reduce nitrogen oxides in automobile exhaust. Here, in addition to Pt and Pd, Rh is also incorporated into the catalyst, due to the fact that in a net reducing feedstream Pt and Pd are severely inhibited by SO_2 while Rh is largely resistant to it and thus is capable of promoting the reduction of nitrogen oxides. (Note again that the components of the catalyst are largely dictated by poisoning considerations.)

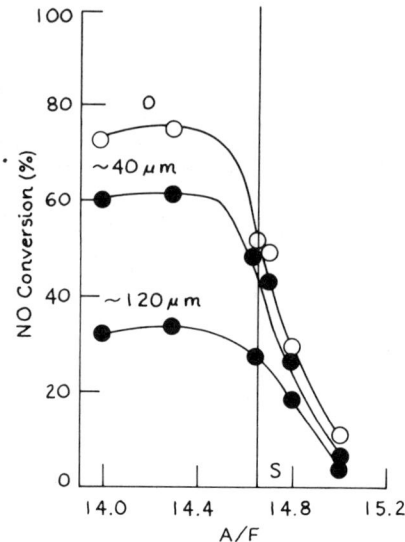

FIG. 60. Effects of Rh location on catalyst performance [222].

Rh is mined as an impurity in the Pt-Pd ore and thus its quantity is limited. Moreover, it is also susceptible to poisoning. Thus, in order to reduce Rh usage, we explored the possibility of impregnating it below the outer surface of the pellets so that it is protected from poisoning. Experiments with Rh bands impregnated to different depths were employed to optimize the thickness of the inert shell around a Rh-impregnabed band, to provide adequate poison resistance without excessive diffusion limitations for the conversion of nitrogen oxides (Fig. 60). Pt was then added to the outer shell, and accelerated poisoning experiments proved the viability of the concept (Fig. 61). Pd was employed in a third layer below Rh, and the resulting catalysts (Hegedus and Summers [248], Summers and Hegedus [249]) thus pointed the way toward reduced Rh usage and improved catalyst performance.

In principle, the pore structure of catalyst pellets should be amenable to formal optimization, since it represents a true compromise between the opposing requirements of high surface area and high diffusivity. Such a study was carried out recently by Hegedus [250], who used a constrained nonlinear optimization technique to arrive at pore structures which favor high activity after a specified time on stream during pore-mouth poisoning.

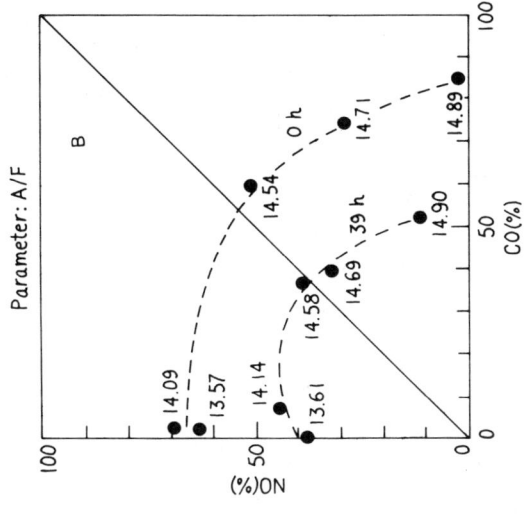
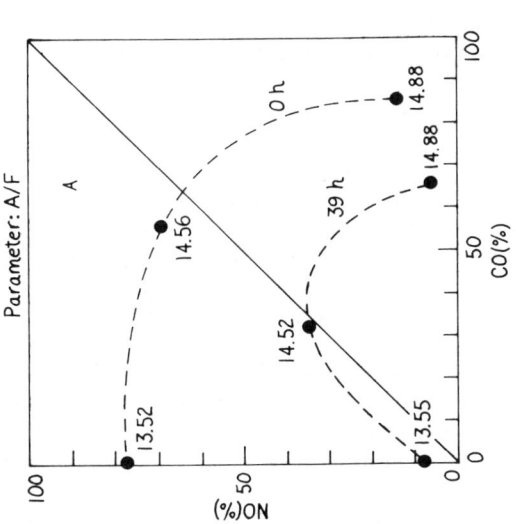

FIG. 61. Comparison of coimpregnated (A) and separately impregnated (Pt external, Rh internal, B) catalysts before and after accelerated poisoning [222].

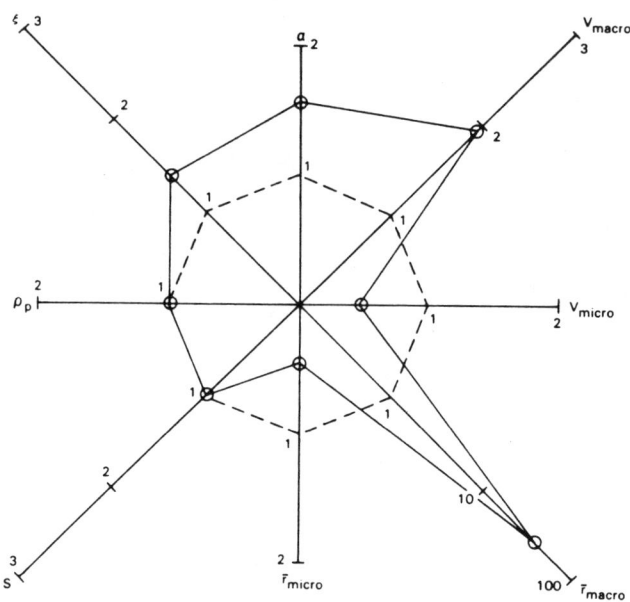

FIG. 62. Snowflake diagram of the pore structure and performance of an optimized catalyst pellet after poisoning in an automobile exhaust converter [250]. See text for details.

The bimodal pore structure of the catalyst pellets was represented by the macro- and micropore volumes, integral-averaged macro- and micropore radii, and by the skeletal density of the alumina. From these, the surface area and pellet density can be directly calculated. Constraints were placed on all of the above parameters during the calculations, in order to restrict them to physically attainable and practical regimes.

The diffusivity of the main reactant and of the poison precursor were computed from the random pore model of Wakao and Smith [251]. From these, the rate of progression of the pore-mouth-poisoned front could be calculated from the Carberry-Gorring model [174]. The diffusion-limited activity of the catalyst pellets was calculated at a fixed poison exposure (i.e., a fixed time period and fixed inlet poison concentration); this, when integrated for the whole reactor, served as the objective function (to be maximized) in the optimization calculations.

CASE HISTORY

Figure 62 shows the results of the calculations with one particular constraint strategy. α is a measure of the activity of the catalyst pellets after a fixed poison exposure, s is the surface area, ρ_p is the pellet density, and ξ is the fractional poison penetration depth. In Fig. 62, the parameters were normalized against the properties of a typical commercial automobile exhaust catalyst: $V_{macro} = 0.366$ cm^3/g, $V_{micro} = 0.627$ cm^3/g, $\bar{r}_{macro} = 6962$ Å, $\bar{r}_{micro} = 110.5$ Å, s = 114.5 m^2/g, $\rho_p = 0.783$ g/cm^3, $\xi = 0.0396$, and $\alpha = 2.69$. The optimization calculations showed that the pellets can, in principle, be reconfigured to provide a 1.6-fold increase in activity after poison exposure (with respect to the reference catalyst), by increasing V_{macro} and $\bar{r}_{macro}$ and decreasing V_{micro} and $\bar{r}_{micro}$, while keeping the same surface area (s) and pellet density (ρ_p). Further improvements were shown to be theoretically possible if pore structures are allowed which result in higher surface areas and lower pellet densities. It would be fascinating to see if the advantages of these "synthetic" pore structures could be verified by appropriate experiments.

References

[1] S. Berkman, J. C. Morrell, and G. Egloff, Catalysis, Reinhold, New York, 1940.
[2] E. B. Maxted, Adv. Catal., 3, 129 (1951).
[3] W. B. Innes, in P. H. Emmett (ed.), Catalysis, Vol. 1, Reinhold, New York, 1954, p. 245.
[4] J. Petro, in Kontakt Katalizis (Z. G. Szabo, ed.), Akademiai Kiado, Budapest, 1966.
[5] J. B. Butt, Adv. Chem., 109, 259 (1972).
[6] J. Petro, in Contact Catalysis (Z. G. Szabo and D. Kallo, eds.), Elsevier, New York, 1976.
[7] H. Knozinger, Adv. Catal., 25, 184 (1976).
[8] J. D. Butt and R. M. Billimoria, ACS Symp. Ser., 72, 288 (1978).
[9] M. J. Oudar, Cat. Rev.—Sci. Eng., 22, 171 (1980).
[10] J. B. Butt, International Symposium on Catalyst Deactivation, Antwerpen, Belgium, October 1980.
[11] J. G. McCarty and H. Wise, J. Chem. Phys., 72, 6332 (1980).
[12] I. Alstrup, J. R. Rostrup-Nielsen, and S. Roen, Applied Catalysis, 1, 303 (1981).

[13] J. Benard, J. Oudar, N. Barbouth, E. Margot, and Y. Berthier, Surf. Sci., 88, L35 (1979).
[14] E. F. G. Herington and E. K. Rideal, Trans. Faraday Soc., 40, 505 (1944).
[15] A. Frennet, G. Lienard, A. Crucq, and L. Degols, J. Catal., 53, 150 (1978).
[16] A. Verma and D. M. Ruthven, Ibid., 46, 160 (1977).
[17] J. A. Schwarz, Surf. Sci., 87, 525 (1979).
[18] P. K. Agrawal, J. R. Katzer, and W. H. Manogue, J. Catalysis, 69, 327 (1981).
[19] Y. Berthier, M. Perdereaux, and J. Oudar, Surf. Sci., 36, 225 (1973).
[20] W. Heegemann, K. H. Meister, E. Berthold, and K. Hayek, Ibid., 49, 161 (1975).
[21] T. E. Fischer and S. R. Kelemen, J. Catal., 53, 24 (1978).
[22] S. Mroz, Surf. Sci., 83, L625 (1979).
[23] W. Erley and H. Wagner, J. Catal., 53, 287 (1978).
[24] H. P. Bonzel and R. Ku, J. Chem. Phys., 58, 4617 (1973).
[25] H. P. Bonzel and R. Ku, Ibid., 59, 1641 (1973).
[26] E. E. Wolf and E. E. Petersen, J. Catal., 47, 28 (1977).
[27] S. Johnson and R. J. Madix, Surf. Sci., 103, 361 (1981).
[28] T. N. Rhodin and C. F. Brucker, Solid State Commun., 23, 275 (1977).
[29] C. H. Rochester and R. J. Terrell, J. Chem. Soc., Faraday Trans. 1, 73, 609 (1977).
[30] D. W. Goodman and M. Kiskinova, Surf. Sci., 105, L265 (1981).
[31] M. Kiskinova and D. W. Goodman, Surf. Sci., 108, 65 (1981).
[32] J. R. Rostrup-Nielsen and K. Pedersen, J. Catalysis, 59, 395 (1979).
[33] E. D. Williams, C. M. Chan, and W. H. Weinberg, Surf. Sci., 81, L309 (1979).
[34] S. R. Kelemen, T. E. Fischer, and J. A. Schwarz, Ibid., 81, 440 (1979).
[35] G. B. Fisher, Ibid., 87, 215 (1979).
[36] C. C. Chang, General Motors Res. Publ. GMR-2781, August 1978.
[37] H. Windawi and J. R. Katzer, Surf. Sci., 75, L761 (1978).
[38] A. W. Cramb, W. R. Graham, and G. R. Belton, Metall. Trans., 9B, 623 (1978).
[39] F. T. Bain, S. D. Jackson, S. J. Thomson, G. Webb, and E. Willocks, J. Chem. Soc., Faraday Trans. 1, 72, 2516 (1976).

REFERENCES

[40] M. J. Baird, D. T. Weinberger, G. Delzer, A. P. Hobbs, P. Pantages, and F. W. Steffgen, Pittsburgh Energy Research Center RI-78/2, 1978.

[41] K. Otto, L. Bertosiewicz, and M. Shelef, Carbon, $\underline{17}$, 351 (1979).

[42] R. B. Pannell, K. S. Chung, and C. H. Bartholomew, J. Catal., $\underline{46}$, 340 (1977).

[43] C. F. Ng and G. A. Martin, Ibid., $\underline{54}$, 384 (1978).

[44] B. J. Wood, W. E. Isakson, and H. Wise, Ind. Eng. Chem., Prod. Res. Dev., In Press.

[45] R. T. Rewick and H. Wise, J. Phys. Chem., $\underline{82}$, 751 (1978).

[46] L. Gonzalez-Tejuca and J. Turkevich, J. Chem. Soc., Faraday Trans. 1, $\underline{74}$, 1064 (1978).

[47] K. Schwaha, N. D. Spencer, and R. M. Lambert, Surf. Sci., $\underline{81}$, 273 (1979).

[48] R. J. Madon and H. Shaw, Catal. Rev.—Sci. Eng., $\underline{15}$, 69 (1977).

[49] M. I. Yanovskii and A. D. Berman, J. Chromatogr., $\underline{69}$, 3 (1972).

[50] M. Shelef, K. Otto, and N. C. Otto, Adv. Catal., $\underline{27}$, 311 (1978).

[51] G. J. K. Acres, B. J. Cooper, E. Shutt, and B. W. Malerbi, Adv. Chem. Ser., $\underline{143}$, 54 (1974).

[52] E. B. Maxted and M. S. Biggs, J. Chem. Soc., p. 3844 (1957)

[53] J. Tsai, P. K. Agrawal, D. R. Sullivan, J. R. Katzer, and W. H. Manogue, J. Catal., $\underline{61}$, 204 (1980).

[54] R. A. Dalla Betta and M. Shelef, Prepr. Div. Fuel Chem., ACS, $\underline{21}$, 43 (1976).

[55] E. B. Maxted and G. T. Ball, J. Chem. Soc., p. 2778 (1954).

[56] E. B. Maxted and G. T. Ball, Ibid., p. 3153 (1953).

[57] L. E. Trimble, Mater. Res. Bull., $\underline{9}$, 1405 (1974).

[58] T. E. Madey, D. W. Goodman, and R. D. Kelley, J. Vac. Sci. Technol., $\underline{16}$, 433 (1979).

[59] Y. Fuji and J. C. Bailar, J. Catal., $\underline{52}$, 342 (1978).

[60] L. L. C. Sorensen and K. Nobe, Ind. Eng. Chem., Prod. Res. Dev., $\underline{11}$, 423 (1972).

[61] L. L. C. Sorensen and K. Nobe, Environ. Sci. Technol., $\underline{6}$, 239 (1972).

[62] G. D. Parks, A. M. Schaffer, M. J. Dreiling, and C. M. Shiblom, Prepr. Div. Pet. Chem., ACS, $\underline{25}$, 335 (1980).

[63] J. R. Katzer and R. Sivasubramanian, Catal. Rev.—Sci. Eng., $\underline{20}$, 155 (1979).

[64] I. Iida and K. Tamaru, J. Catal., $\underline{56}$, 229 (1979).

REFERENCES

[65] F. L. Williams and K. Baron, Ibid., 40, 108 (1975).
[66] J. Tsai, P. K. Agrawal, J. M. Foley, J. R. Katzer, and W. H. Manogue, Ibid., 61, 192 (1980).
[67] R. D. Kelley, D. W. Goodman, T. E. Madey, and J. T. Yates, Conference on Catalyst Deactivation and Poisoning, Lawrence Berkeley Laboratory, Berkeley, California, May 1978.
[68] S. Affrossman, T. Donnelly, and J. McGeachy, J. Catal., 29, 346 (1973).
[69] A. Frackiewicz, Z. Karpinski, A. Leszczynski, and W. Palczewska, Proceedings of the Fifth International Congress on Catalysis, North-Holland, Amsterdam, 1973, p. 635.
[70] L. D. Schmidt and D. Luss, J. Catal., 22, 269 (1971).
[71] G. A. Somorjai, Ibid., 27, 453 (1972).
[72] J. J. McCaroll, T. Edmonds, and R. C. Pitkethly, Nature, 223, 1260 (1969).
[73] C. Herring, Phys. Rev., 82, 87 (1951).
[74] T. D. Halachev and E. Ruckenstein, Surf. Sci., 108, 292 (1981).
[75] R. W. McCabe, T. Pignet, and L. D. Schmidt, J. Catal., 32, 114 (1974).
[76] M. Flytzani-Stephanopoulos, S. Wong, and L. D. Schmidt, Ibid., 49, 51 (1977).
[77] H. Amariglio and G. Rambeau, Proceedings of the Sixth International Congress on Catalysis, The Chemical Society, London, 1977, p. 1113.
[78] D. W. Blakely and G. A. Somorjai, Surf. Sci., 65, 419 (1977).
[79] N. A. Fishel, R. K. Lee, and F. C. Wilhelm, Environ. Sci. Technol., 8, 260 (1974).
[80] T. R. Ingraham, Trans. Metall. Soc., AIME, 233, 359 (1965).
[81] P. S. Lowell, K. Schwitzgebel, T. B. Parsons, and K. J. Sladek, Ind. Eng. Chem., Proc. Res. Dev., 10, 384 (1971).
[82] H. D. Simpson, Adv. Chem. Ser., 143, 39 (1975).
[83] A. Amirnazmi and M. Boudart, J. Catal., 39, 383 (1975).
[84] B. G. Baker and R. Peterson, Proceedings of the Sixth International Congress on Catalysis, The Chemical Society, London, 1977, p. 988.
[85] Y. Kim, S. K. Shi, and J. M. White, J. Catal., 61, 374 (1980).
[86] B. J. Wood, H. Wise, and R. S. Yolles, Ibid., 15, 355 (1969).
[87] H. S. Taylor, Proc. R. Soc., A108, 105 (1925).
[88] F. H. Constable, Ibid., A108, 355 (1925).
[89] J. Volter and M. Hermann, Z. Anorg. Allg. Chem., 405, 315 (1974).
[90] G. Schulz-Ekloff, D. Baresel, and W. Sarholz, J. Catal., 43, 353 (1976).

[91]   K. Baron, Thin Solid Films, 55, 449 (1978).
[92]   R. D. Clay and E. E. Petersen, J. Catal., 16, 32 (1970).
[93]   P. N. Ross and P. Stonehart, Ibid., 35, 391 (1974).
[94]   J. G. Larson and W. K. Hall, J. Phys. Chem., 69, 3080 (1965).
[95]   N. Ray, V. K. Rastogi, H. Mahapatra, and S. P. Sen, J. Res. Inst. Catal., Hokkaido Univ., 21, 187 (1973).
[96]   R. A. Dalla Betta, A. G. Piken, and M. Shelef, J. Catal., 40, 173 (1975).
[97]   Y. Inoue, I. Kojima, S. Moriki, and I. Yasumori, Proceedings of the Sixth International Congress on Catalysis, The Chemical Society, London, 1977, p. 139.
[98]   S. Szepe and O. Levenspiel, Proceedings of the Fourth European Symposium on Chemical Reaction Engineering, Pergamon, Oxford, 1971, p. 265.
[99]   J. B. Butt, C. K. Wachter, and R. M. Billimoria, Chem. Eng. Sci., 33, 1321 (1978).
[100] J. B. Butt, Proceedings of the Fourth European Symposium on Chemical Reaction Engineering, Pergamon, Oxford, 1971, p. 255.
[101] J. E. Dabrowski, J. B. Butt, and H. Bliss, J. Catal., 18, 297 (1970).
[102] P. C. Saunders and J. W. Hightower, Prepr., Div. Pet. Chem., ACS, 15, A79 (1970).
[103] M. P. Rosynek and J. W. Hightower, Proceedings of the Fifth International Congress on Catalysis, North-Holland, Amsterdam, 1973, p. 851.
[104] T. T. Chuang, I. G. Dalla Lana, and C. L. Liu, J. Chem. Soc., Faraday Trans. 1, 69, 643 (1973).
[105] A. Ghorbel, C. H.-Van, and S. J. Teichner, J. Catal., 33, 123 (1974).
[106] A. J. Van Roosmalen, M. C. G. Hartmann, and J. C. Mol, J. Catalysis, 66, 112 (1980).
[107] H. Knozinger, H. Kritenbrink, H.-D. Muller, and W. Schulz, Proceedings of the Sixth International Congress on Catalysis, The Chemical Society, London, 1977, p. 183.
[108] I. Mochida, A. Uchino, H. Fujitsu, and K. Takeshita, J. Catal., 43, 264 (1976).
[109] M. P. Rosynek and F. L. Strey, Ibid., 41, 312 (1976).
[110] V. S. Hariharakrishnan, N. Venkatasubramanian, and C. N. Pillai, Ibid., 53, 232 (1978).
[111] J. Take, T. Ueda, and Y. Yoneda, Bull. Chem. Soc. Jpn., 51, 1581 (1978).

[112] K. Mizuno, M. Ikeda, T. Imokawa, J. Take, and Y. Yoneda, Ibid., 49, 1788 (1976).
[113] K. R. Bakshi and G. R. Gavalas, AIChE J., 21, 494 (1975).
[114] K. R. Bakshi and G. R. Gavalas, J. Catal., 38, 326 (1975).
[115] G. P. Lozos and B. M. Hoffman, J. Phys. Chem., 78, 2110 (1974).
[116] V. H. Bremer, K.-H. Steinberg, and T.-K. Chuong, Z. Anorg. Allg. Chem., 403, 72 (1974).
[117] Y. S. Khodakov, P. A. Makarov, G. Delzer, and K. M. Minachev, J. Catal., 61, 184 (1980).
[118] H. Hattori and A. Satoh, Ibid., 45, 32 (1976).
[119] M. Hattori, K. Maruyama, and K. Tanabe, Ibid., 44, 50 (1976).
[120] T. A. Gilmore and J. J. Rooney, J. Chem. Soc., Chem. Commun., p. 219 (1975).
[121] D. D. Eley, C. H. Rochester, and M. S. Scurrell, J. Chem. Soc., Faraday Trans. 1, 69, 660 (1973).
[122] G. Webb and J. I. Macnab, J. Catal., 26, 226 (1972).
[123] R. L. Burnett and T. R. Hughes, Ibid., 31, 55 (1973).
[124] M. J. Sterba and V. Haensel, Ind. Eng. Chem., Prod. Res. Dev., 15, 2 (1976).
[125] P. G. Menon and J. Prasad, Proceedings of the Sixth International Congress on Catalysis, The Chemical Society, London, 1977, p. 1061.
[126] A. V. Ramaswamy, P. Ratnasamy, S. Sivasanker, and A. J. Leonard, Ibid., p. 855.
[127] A. A. Olsthoorn and C. Baelhouwer, J. Catal., 44, 207 (1976).
[128] M. Lajacona, J. L. Verbeek, and G. C. Schuit, Proceedings of the Fifth International Congress on Catalysis, North-Holland, Amsterdam, 1973, p. 103.
[129] H. C. Lee and J. B. Butt, J. Catal., 49, 320 (1977).
[130] R. F. Howe and C. Kemball, J. Chem. Soc., Faraday Trans. 1, 70, 1153 (1974).
[131] J. M. Balois and J. P. Beaufils, React. Kin. Catal. Lett., 3, 355 (1975).
[132] S. W. Cowley and F. E. Massoth, J. Catal., 51, 291 (1978).
[133] W. S. Millman and W. K. Hall, J. Phys. Chem., 83, 427 (1979).
[134] P. Ratnasamy, A. V. Ramaswamy, and S. Sivasanker, J. Catal., 61, 519 (1980).
[135] R. B. Shalvoy and P. J. Reucroft, J. Vac. Sci. Technol., 16, 567 (1979).
[136] P. W. Wentrcek, J. G. McCarty, C. M. Ablow, and H. Wise, J. Catal., 61, 232 (1980).

REFERENCES

[137] H. S. Gandhi, H. C. Yao, H. K. Stepien, and M. Shelef, Society of Automotive Engineers, Paper No. 780606, Detroit, June 1978.
[138] J. C. Summers and K. Baron, J. Catal., 57, 380 (1979).
[139] C. Kemball, Catalysis, Prog. Res., Proc. NATO Sci. Comm. Conf., 1973, p. 85.
[140] J. Mooi, J. P. Kuebrich, M. F. L. Johnson, and F. J. Chloupek, Proc. Div. Refin., Am. Pet. Inst., 53, 14 (1973).
[141] A. G. Moldovan, A. Elattar, and W. E. Wallace, J. Solid State Chem., 25, 23 (1978).
[142] D. W. Johnson, P. K. Gallagher, E. M. Vogel, and F. Schrey, Proceedings of the Fourth International Conference on Thermal Analysis, Budapest, Vol., 3, 1974, p. 181.
[143] P. K. Gallagher, D. W. Johnson, E. M. Vogel, and F. Schrey, Mater. Res. Bull., 10, 623 (1975).
[144] Y. F. Y. Yao, Am. Ceram. Soc. Bull., 53, 342 (1974).
[145] J. Barbier, A. Morales, P. Marecot, and R. Mauerel, Bull. Soc. Chim. Belg., 88, 569 (1979).
[146] W. H. Manogue and J. R. Katzer, J. Catal., 32, 166 (1974).
[147] M. Boudart, A. Aldag, J. E. Benson, N. A. Dougharty, and C. G. Harkins, Ibid., 6, 92 (1966).
[148] I. Iida and K. Tamaru, Z. Phys. Chem., N.F. 107, 231 (1977).
[149] T. A. Dorling, M. J. Eastlake, and R. L. Moss, J. Catal., 14, 23 (1969).
[150] L. L. Hegedus and E. E. Petersen, Ibid., 28, 150 (1973).
[151] M. Boudart, A. W. Aldag, and M. A. Vannice, Ibid., 18, 46 (1970).
[152] S. Fuentes and F. Figueras, Ibid., 54, 397 (1978).
[153] J. J. Ostermaier, J. R. Katzer, and W. H. Manogue, Ibid., 41, 277 (1976).
[154] R. W. Joyner, J. Chem. Soc., Faraday Trans. 1, 76, 357 (1980).
[155] P. Gallezot, J. Datka, J. Massardier, M. Primet, and B. Imelik, Proceedings of the Sixth International Congress on Catalysis, The Chemical Society, London, 1977, p. 696.
[156] A. Wheeler, Adv. Catal., 3, 249 (1951).
[157] L. L. Hegedus and E. E. Petersen, Catal. Rev.—Sci. Eng., 9, 245 (1974).
[158] E. E. Petersen, in Experimental Methods in Catalytic Research, Vol. 2 (R. B. Anderson, ed.), Academic, New York, 1975.
[159] J. R. Balder and E. E. Petersen, J. Catal., 11, 202 (1968).

[160] J. R. Balder and E. E. Petersen, Chem. Eng. Sci., 23, 1287 (1968).
[161] N. A. Dougharty, Ibid., 25, 489 (1970).
[162] L. L. Hegedus and E. E. Petersen, Ibid., 28, 69 (1973).
[163] L. L. Hegedus and E. E. Petersen, Ind. Eng. Chem., Fundam., 11, 579 (1972).
[164] B. A. Tennant and J. Wei, American Institute of Chemical Engineers, New York, December 1977, Paper 25a.
[165] R. K. Herz, General Motors Research Laboratories, Personal Communication.
[166] J. L. Hahn and E. E. Petersen, Can. J. Chem. Eng., 48, 147 (1970).
[167] L. L. Hegedus and E. E. Petersen, Chem. Eng. Sci., 28, 345 (1973).
[168] E. E. Wolf and E. E. Petersen, J. Catal., 46, 190 (1977).
[169] E. E. Wolf and E. E. Petersen, Chem. Eng. Sci., 29, 1500 (1974).
[170] E. E. Wolf and E. E. Petersen, Ibid., 32, 493 (1977).
[171] F. Gioia, F. Alfani, and G. Greco, Ibid., 27, 1745 (1972).
[172] K. B. Bischoff, Ibid., 18, 711 (1963).
[173] S. Masamune and J. M. Smith, AIChE J., 12, 384 (1966).
[174] J. J. Carberry and R. L. Gorring, J. Catal., 5, 529 (1966).
[175] J. H. Olson, Ind. Eng. Chem., Fundam., 7, 185 (1968).
[176] C. Chu, Ibid., 7, 509 (1968).
[177] Y. Murakami, T. Kobayashi, T. Hattori, and M. Masuda, Ibid., 7, 599 (1968).
[178] F. Gioia, L. G. Gibilaro, and G. Greco, Chem. Eng. J., 1, 9 (1970).
[179] F. Gioia, Ind. Eng. Chem., Fundam., 10, 204 (1971).
[180] R. P. Merrill, J. Catal., 50, 184 (1977).
[181] L. L. Hegedus, Ind. Eng. Chem., Fundam., 13, 190 (1974).
[182] J. J. Zwicky and G. Gut, Chem. Eng. Sci., 33, 1363 (1978).
[183] E. K. T. Kam, P. A. Ramachandran, and R. Hughes, J. Catal., 38, 283 (1975).
[184] B. Valdman, P. A. Ramachandran, and R. Hughes, Ibid., 42, 303 (1976).
[185] L. J. Christiansen and S. L. Anderson, Chem. Eng. Sci., 35, 314 (1980).
[186] S. T. Lee and R. Aris, ACS Symp. Ser., 65, 110 (1978).
[187] R. Tartarelli and F. Morelli, J. Catal., 11, 159 (1968).
[188] V. Hlubucek and J. Pasek, Int. Chem. Eng., 12, 270 (1972).
[189] N. M. Tai and P. F. Greenfield, Chem. Eng. J., 16, 89 (1978).
[190] B. D. Kulkarni and P. A. Ramachandran, Ibid., 19, 57 (1980).

[191] M. P. Dudukovic, AIChE J., 22, 945 (1976).
[192] S. J. Khang and O. Levenspiel, Ind. Eng. Chem., Fundam., 12, 185 (1973).
[193] L. L. Hegedus, Chem. Eng. Sci., 29, 2003 (1974).
[194] L. L. Hegedus, S. H. Oh, and K. Baron, AIChE J., 23, 632 (1977).
[195] E. G. Schlosser, Chem.-Ing. Tech., 47, 997 (1975).
[196] R. B. Anderson, F. S. Karn, and J. F. Schultz, J. Catal., 4, 56 (1965).
[197] F. Moseley, R. W. Stephens, K. D. Stewart, and J. Wood, Ibid., 24, 18 (1972).
[198] M. El Menshawy, E. Schutt, and U. Wiesmann, Chem.-Ing. Tech., 46, 1059 (1974).
[199] E. Sada and C. Y. Wen, Chem. Eng. Sci., 22, 559 (1967).
[200] A. N. Gokarn and L. K. Doraiswamy, Ibid., 28, 401 (1973).
[201] R. Aris, Ibid., 6, 262 (1957).
[202] K. Rajagopalan and D. Luss, Ind. Eng. Chem., Proc. Des. Dev., 18, 459 (1979).
[203] J. T. Richardson, Ibid., 11, 12 (1972).
[204] H. C. Chen and R. B. Anderson, J. Catal., 28, 161 (1973).
[205] J. L. Bomback, M. A. Wheeler, J. Tabock, and J. D. Janowski, Environ. Sci. Technol., 9, 139 (1975).
[206] T. S. Chou and L. L. Hegedus, AIChE J., 24, 255 (1978).
[207] J. J. Stanulonis, B. C. Gates, and J. H. Olson, Ibid., 22, 576 (1976).
[208] E. Newson, Ind. Eng. Chem., Proc. Des. Dev., 14, 27 (1975).
[209] G. P. Androutsopoulos and R. Mann, Chem. Eng. Sci., 33, 673 (1978).
[210] P. A. Ramachandran and J. M. Smith, AIChE J., 23, 353 (1977).
[211] L. L. Hegedus, T. S. Chou, J. C. Summers, and N. M. Potter, in Preparation of Catalysts II (B. Delmon, P. Grange, P. Jacobs, and G. Poncelet, eds.), Elsevier, Amsterdam, 1979.
[212] E. Michalko, U.S. Patents 3,259,454 and 3,259,589 (both July 5, 1966).
[213] J. Hoekstra, U.S. Patent 3,388,077 (June 11, 1968).
[214] F. Shadman-Yazdi and E. E. Petersen, Chem. Eng. Sci., 27, 227 (1972).
[215] W. E. Corbett and D. Luss, Ibid., 29, 1473 (1974).
[216] G. B. DeLancey, Ibid., 28, 105 (1973).
[217] J. Wei and E. R. Becker, Adv. Chem. Ser., 143, 166 (1975).
[218] S. Minhas and J. J. Carberry, J. Catal., 14, 270 (1969).
[219] E. R. Becker and J. Wei, Ibid., 46, 372 (1977).

[220] L. L. Hegedus and J. C. Summers, Ibid., 48, 345 (1977).
[221] J. C. Summers and L. L. Hegedus, Ibid., 51, 185 (1978).
[222] L. L. Hegedus, J. C. Summers, J. C. Schlatter, and K. Baron, Ibid., 56, 321 (1979).
[223] E. E. Wolf, Ibid., 47, 85 (1977).
[224] R. E. Polomski and E. E. Wolf, Ibid., 52, 272 (1978).
[225] E. E. Wolf, AIChE J., 26, 55 (1980).
[226] S. H. Oh, J. C. Cavendish, and L. L. Hegedus, AIChE Journal, 26, 935 (1980).
[227] H. Komiyama and H. Inoue, J. Chem. Eng. Jpn., 3, 206 (1970).
[228] N. G. Karanth and D. Luss, Chem. Eng. Sci., 30, 695 (1975).
[229] T. J. Pareja and D. Luss, Ibid., 30, 1219 (1975).
[230] J. W. Lee and J. B. Butt, Chem. Eng. J., 6, 111 (1973).
[231] J. J. Carberry, Ind. Eng. Chem., 58, 40 (1966).
[232] M. Sagara, S. Masamune, and J. M. Smith, AIChE J., 13, 1226 (1967).
[233] W. H. Ray, Chem. Eng. Sci., 27, 489 (1972).
[234] H. P. Koh and R. Hughes, AIChE J., 20, 395 (1974).
[235] J. B. Butt, D. M. Downing, and J. W. Lee, Ind. Eng. Chem., Fundam., 16, 270 (1977).
[236] J. J. Carberry, Ibid., 14, 129 (1975).
[237] S. H. Oh, L. L. Hegedus, and R. Aris, Ibid., 4, 309 (1978).
[238] H. H. Lee, Chem. Eng. Sci., 35, 905 (1980).
[239] J. W. Lee, J. B. Butt, and D. M. Downing, AIChE J., 24, 212 (1978).
[240] D. M. Downing, J. W. Lee, and J. B. Butt, Ibid., 25, 461 (1979).
[241] D. P. Burke, Chem. Week, p. 42 (March 28, 1979).
[242] L. L. Hegedus and J. J. Gumbleton, ChemTech, 10, 630 (1980).
[243] L. L. Hegedus and K. Baron, J. Catal., 54, 155 (1978).
[244] L. L. Hegedus and J. C. Summers, U.S. Patent 4,119,571 (October 10, 1978).
[245] J. R. Adomaitis, J. E. Smith, and D. E. Achey, Society of Automotive Engineers, Detroit, Michigan, February 15, 1980, Paper No. 800084.
[246] M. Chen and L. D. Schmidt, J. Catal., 56, 198 (1979).
[247] J. C. Summers and L. L. Hegedus, U.S. Patent 4,152,301 (May 1, 1979).
[248] L. L. Hegedus and J. C. Summers, U.S. Patent 4,128,506 (December 5, 1978).
[249] J. C. Summers and L. L. Hegedus, U.S. Patent 4,153,579 (May 8, 1979).
[250] See also L. L. Hegedus, Ind. Eng. Chem., Prod. Res. Dev., 19, 533 (1980).
[251] J. M. Smith, "Chemical Engineering Kinetics," McGraw-Hill, New York, 1970.

Index

A

Acetic acid, 36
Acetylene, 18-19, 33
Acrolein, 28
Action radius, 7
Active site, 29, 48
Adsorption
  competitive, 1, 9-10
  irreversible, 1, 41
  Langmuirian, 6
  multisite, 10, 29
  on catalyst support, 43
  reversible, 1, 41
  single-site, 9-10
Alloy formation, 21, 22, 93
Alumina, 35, 36, 43, 82
Aluminum ion, 35-38
Aluminum-o-phosphate, 43
Ammonia, 20-21, 22, 35, 36, 42, 44-45, 46
Ammonia oxidation, 22, 23, 46
Ammonia synthesis, 23, 31, 67
Antimony, 21
Arsenic, 40
Arsenic hydride, 31
Automobile exhaust catalysis, 20, 22, 24, 42, 43, 67, 81 ff

B

Barium oxide, 37
Base metal catalysts, 26
Benzene, 18-19, 22, 33
Bifunctional catalysts, 34
Bimetallic catalysts, 40
Binding states, 13
Biot number, 63, 85, 86, 88
Bonding, 19
Bronsted acid sites, 35
n-Butylamine, 38

C

"Cage" model, 19
Carbon, 19, 22
Carbon dioxide, 35, 36
Carbon monoxide, 12-16, 19-20, 23-24, 30-31, 35, 43-44, 63, 65, 69
Catalytic etching, 23
Catalysts, types of
  monofunctional, 5, 28
  multi-functional, 5

Catalyst sites, types of
  distributed, 5
  monofunctional, 5
Chemisorption
  carbon monoxide, 41
  multisite, 7
  single-site, 9-10
  sulfur, 6, 13-14
Chromia, 37
Chromium, 22
Cobalt-molybdenum on alumina, 41
Coking, 2, 21, 40, 41, 46, 67
Compound formation, 5, 23-28
Copper-chromium catalyst, 24
Cordierite, 43
COS, 36
Cumene cracking, 20
Cyclohexene, 20
Cyclopropane, 31

D

Deactivation of catalysts
  chemical, 1
  mechanical, 1
  thermal, 1
Dehydration
  of γ-alumina, 35
  of methanol, 38
  of ethanol, 38
Dehydrocyclization, 41
Dehydrogenation
  acetylene, 18-19
  benzene, 18-19
  methylcyclohexane, 56
Dehydroisomerization, 41
Demetallation, 66
Diffusion, 29, 47
Diffusivity measurements, 67

Dimethylphenylarsine, 21
Dinitrogen oxide, 27
Dispersion of metals, 45
Displacement of poison species, 21
Disproportionation of alkanes, 40
Dissolution, 6, 21-22

E

Electronic effects, 8, 18
Electron microprobe, 83
Eley-Rideal mechanism, 16
Epimerization of cis-1,2-dimethyl cyclohexane, 44
Ethanol, 38
Ethylene, 21, 22, 23, 33
Exchange reactions
  D_2 with 1-butene, 39, 40
  D_2 with olefins, 35-36, 39
  D_2 with H_2O, 46
Exothermic reactions, 79, 80

F

Free energy of surfaces, 22-23
Fischer-Tropsch synthesis, 20

G

Gallium oxide, 37
Graphite, 21, 44

INDEX

H

Halogens, 83
Hydrocracking, 41
Hydrodenitrogenation, 21
Hydrodesulfurization, 67
Hydrogenation
  acetylene, 33, 76
  benzene, 44, 46, 80
  1-butene, 39, 40
  cyclohexene, 20
  cyclopropane, 45, 46, 55
  ethylene, 22, 33, 46, 79
  isoprene, 21
Hydrogenolysis
  cyclopentane, 44-45
  cyclopropane, 31
Hydrogen sulfide, 6, 13, 15,
  21-22, 31, 32, 43-44
Hydrophobic support, 46
Hydroxyl radicals, 35

I

Impregnation profiles, 36,
  68-71, 77, 88, 92
  94
Infrared spectroscopy, 19, 36,
  44, 53
Iron carbonyl, 21, 65
Iron oxide, 27
Islands of adsorbed species,
  15-16, 46
Isomerization
  1-butene, 36, 39, 40
  cyclopropane, 33
  olefins, 35, 36
Isoprene, 21

L

Langmuir adsorption isotherm,
  6, 58
Langmuir-Hinshelwood rate mechanism, 17, 19, 76, 77
$La_{0.7}$, $Pb_{0.3}$, MnO_3, 44
Lanthana, 37
Lead, 20-22, 31, 43-44, 81, 82,
  83
Lead aluminate, 43
Lead aluminum silicate, 43
Lead sulfate, 43
Lewis acid sites, 35
Lubricating oil, 82

M

Mass spectrometry, 53
Mass transport
  external phase, 61
  fluid phase, 47
  interparticle, 47, 48
  intraparticle, 47
  of poison precursors, 47
  surface, 47
Mercury, 22, 39, 40
Metathesis catalyst, 41
Methanation, 21, 22, 32-33, 41,
  43
Methanol, 38
Methylcyclohexane, 56
Mobility of adsorbed species,
  18
Molybdenum on alumina catalyst,
  41
Monofunctional catalysts, 28
Monolith catalysts, 43

Monte Carlo techniques, 9, 19, 35
Mullite, 43
Multifunctional catalysts, 34-39, 42-43, 78
Multiple steady states, 63, 79

N

Nickel, 20-22, 32, 33, 43-44, 76
Nickel on alumina, 6, 41, 43, 44, 65
$NiAlS_4$, 43
$Ni-Cr-MgSiO_3$, 41
Nickel-copper alloy, 22
Nickel-iridium on alumina, 41
Nickel oxide, 27, 65
Nickel on silica, 44
Nickel sulfide, 6, 44
Nickel on zirconia, 44
Nitric oxide, 18, 19, 20, 21, 22, 26, 27, 35
Non-separable kinetics, 38

O

Optimization of catalyst particle size, 65
Oxidation
  ammonia, 22, 23
  carbon monoxide, 14-16, 23, 31, 44, 63, 65, 69, 82
  ethane, 44
  ethylene, 21
  hydrocarbons, 82
  iron, 27
  nickel, 27
  propane, 26

Oxidation (continued)
  propylene, 28
  rhodium, 28
Oxide catalysts, 35
Oxygen, 19, 20

P

Palladium, 21, 22, 33, 71, 82, 90, 92, 93
Particle size, 46, 65
Pellet size, 64
Phenothiazine, 36
Phosphoric acid, 83
Phosphorus, 19, 41, 81-86
Platinum, 11-23, 26, 27, 30-31, 46, 71, 74, 90, 92
Platinum on alumina, 40, 41, 44, 45, 46, 55, 56, 63, 69, 79, 82, 92
Platinum arsenide, 31
Platinum-iridium or alumina, 41
Platinum oxide, 27, 46
Platinum-palladium alloy, 82
Platinum-rhenium on alumina, 40, 41
Platinum on silica, 46
Platinum Y-zeolite, 46
Poison
  catalytically active, 62
  precursor, 1, 20, 83
"Poison-getter," 43, 83
Poisoning, types of
  antiselective, 29, 42, 43, 50, 51
  beneficial, 41
  complete, 10
  core, 51, 52, 61, 76
  impurity, 2, 34, 52, 61, 65, 69, 70, 78

INDEX

Poisoning (continued)
  irreversible, 44, 57, 59
  non-selective, 29, 31-32, 34, 49
  pore mouth, 50, 52, 61, 63, 66, 67, 70, 74-76
  reversible, 44, 57, 58
  selective, 29, 33-40, 50, 71, 75-78
  shell, 83-88
  uniform, 52, 69, 76, 77
Pontryagin's minimum principle, 69
Pore plugging, 67
Pore structure, 66
Presulfiding, 41
Propane oxidation, 23
Propylene oxidation, 28

Q

Quasi-steady-state assumption, 59

R

Radioactive tracer techniques, 39
Raney nickel, 43
Reconstruction of catalyst surfaces, 5, 22-23
Reforming of hydrocarbons, 40, 41
Regeneration of catalysts, 6
Restructuring of catalysts, 1
Reviews of Catalyst Poisoning, 3
Rhenium oxide, 41
Rhodium, 32, 43, 71, 93, 94
Ruthenium, 20, 32, 43
Ruthenium on alumina, 41

S

Segregation in alloys, 2
Selectivity, 28-32
Separable kinetics, 34
Shape
  of catalyst poisoning curves, 47
  of catalyst particles, 66
Silica, 40
Silica-alumina catalyst, 20, 37
Single-pellet diffusion reactor, 51-56
Sintering, 2, 31
Sites, types of
  acidic, 35-39
  basic, 35-39
  electron transfer, 35
Site strength distribution, 28, 30, 33, 44
Spillover, 46
Structure of adsorbed overlayers, 11-13
Structure sensitivity, 44-45
Sulfidation, 41
Sulfide catalysts, 40, 41
Sulfospinels, 43
Sulfur, poisoning of
  automobile exhaust catalysts, 81, 82
  chromium, 7
  copper, 7, 24
  iron, 19
  nickel, 6, 7, 22, 44
  palladium, 22
  platinum, 11-19
  platinum-on-alumina catalyst, 40
  platinum-10% rhodium gauzes, 22
  platinum-Y-zeolite, 46
  ruthenium, 22

Sulfur dioxide, 22, 36, 43, 82,
  83, 93
Sulfur trioxide, 43
Surface analysis techniques
  Auger electron spectroscopy,
    11, 22
  field emission microscopy, 33
  low energy electron diffraction, 11-13
  thermal desorption spectroscopy, 12-13, 35
  ultraviolet photoemission spectroscopy, 19
Surface-to-volume ratio (catalyst pellets), 66

$ThNi_5$, 43
Thorium oxide, 44
Transient poisoning processes,
  56, 57
Tungsten carbide, 31
Tungsten oxide, 40

V

Van der Waals radius, 12
Volatization of metals, 2

T

Temperature gradients, 78-80
Tetracyanoethylene, 36
Tetraethyl lead, 81
Thermodynamics of poisoning
  processes, 6-7, 24, 25
Thiele parameter, 49, 52, 63, 64,
  66, 75, 78
Thionaphthene, 21
Thiophene, 21, 40

W

Water, 21, 35, 45, 46

Z

Zeolites, 3, 46
$ZrNi_5$, 43